WIND EFFECTS ON COMPLIANT OFFSHORE STRUCTURES

Proceedings of a session at
Structures Congress '86
sponsored by the Structural Division of the
American Society of Civil Engineers

Hyatt Regency Hotel
New Orleans, Louisiana
September 15-18, 1986

Edited by Charles E. Smith and Emil Simiu

ASCE
1852

Published by the
American Society of Civil Engineers
345 East 47th Street
New York, New York 10017

ABSTRACT

In recent years modern experimental and analytical techniques for estimating wind effects have increasingly been applied to compliant offshore structures. The six papers included in this book are concerned with various aspects and applications of these techniques. Comparisons are reported between aerodynamic forces and moments measured in the wind tunnel and their counterparts as calculated in accordance with rules issued by classification societies, professional associations, or code authorities. It is noted that in most cases these rules yield conservative results, but that this is not always the case. Various methodologies for obtaining and using wind tunnel data are discussed, and techniques are reported for measuring wind and wave effects simultaneously. It is noted that in the present state of the art laboratory tests cannot achieve the combination of large Reynolds numbers (of the order of one million) and small Keulegan-Carpenter numbers (of the order of unity) that is typical of tension leg platform columns. Such tests can therefore provide incorrect estimates of the hydrodynamic damping inherent in full-scale surge motions and, consequently, of the wind-induced dynamic effects of the prototype structure.

The Society is not responsible for any statements made or opinions expressed in its publications.

No part of this publication may be reproduced, stored in a retrieval system, or transmitted, in any form or by any means, electronic, mechanical, photocopying, recording, or otherwise, without prior written permission of the publisher.

Library of Congress Cataloging-in-Publication Data

Wind effects on compliant offshore structures.

Includes indexes.
Contents: Modeling wind loads on mobile offshore structure/by J.M. Macha—Wind effects on semisubmersibles and other floating offshore structures/by E.T.D. Bjerregaard and S.O. Hansen—Wind loads of offshore drilling platforms/by B.J. Vickery, P. Freazy, and S. Helliwell—[etc.]
    1. Offshore structures—Aerodynamics—Congresses. I. Smith, Charles E., 1942-   . II. Simiu, Emil. III. Structures Congress '86 (1986 : New Orleans, La.) IV. American Society of Civil Engineers. Structural Division.

TC1665.W56 1986   627'.98   86-20611
ISBN 0-87262-555-9

Copyright © 1986 by the American Society of Civil Engineers,
All Rights Reserved.
Library of Congress Catalog Card No.: 86-20611
ISBN 0-87262-555-9
Manufactured in the United States of America.

FOREWORD

In recent years modern experimental and analytical techniques for estimating wind effects have increasingly been applied to compliant offshore structures. The six papers included in these Proceedings are concerned with various aspects and applications of these techniques.

Typical of most offshore platforms are relatively intricate configurations which result in substantial aerodynamic shielding effects as well as in lift forces that can seriously affect the overturning moment of the structure. Comparisons reported in these Proceedings between aerodynamic forces and moments measured in the wind tunnel, and their counterparts estimated in accordance with rules adopted by classification societies, professional associations, or code authorities, show differences that can be significant. Although in most cases these rules yield conservative results, this is not always the case. For the sake of safety and economy, it is therefore desirable to obtain data from properly conducted wind tunnel tests.

Another concern related to the action of wind on compliant offshore platforms is the magnitude of the dynamic effects due to atmospheric turbulence. Wind tunnel tests are useful for estimating the fluctuating wind loads acting on platforms. The extent to which these fluctuating loads can be effective in causing dynamic amplifications of the resonant type is more difficult to ascertain, especially for tension leg platforms. Indeed, in this case reliable information on full-scale hydrodynamic damping forces, which control the magnitude of these amplifications, is not available. Moreover, these forces correspond to large Reynolds numbers and, for any given Keulegan-Carpenter number, differ markedly from their laboratory counterparts.

The papers included in these Proceedings deal with various aspects of wind tunnel testing and with approaches to the use of information obtained in wind tunnel and/or in hydrodynamic tests. Among the topics discussed in the papers are: the simulation of aerodynamic forces on members with circular cross-section, for which Reynolds number requirements are routinely violated in the laboratory; the measurement of the aerodynamic (above-water) and hydrodynamic (below water) lift, drag, and moment for semisubmersible platforms; the effect of the sea-air interface; and the effects of hydrodynamic damping on tension leg platform motions.

In most cases the aerodynamic information obtained in the wind tunnel is used as input in analytical models describing platform behavior. This approach to the use of measured data is implicit in most of the work reported here. However, one of the papers reports on a global physical test, in which a tension leg platform immersed in a wave tank is simultaneously subjected to turbulent winds and random waves. Such tests provide the requisite information on model structural response directly, rather than through the agency of an analytical framework. They can be used for the validation (or otherwise) of the analytical frameworks being proposed. A validated framework used in conjunction with hydrodynamic data applicable to the prototype situation can then be employed to model the full-scale behavior of the platform.

The session "Wind Effects on Compliant Offshore Structures," whose Proceedings are presented in this volume, was organized at the initiative of the Committee on Reliability of Offshore Structures of the Structural Division, ASCE, with the encouragement and support of the Minerals Management Service, United States Department of the Interior. Prior to publication, papers were reviewed by a committee charred by the editors. All papers are eligible for discussion in the Journal of Structural Engineering and all papers are eligible for ASCE awards.

Emil Simiu
Chairman, Subcommittee on Environmental Forces
Committee on Reliability of Offshore Structures, ASCE

CONTENTS

Modeling Wind Loads on Mobile Offshore Structure—A Summary of Wind Tunnel Results
  J. M. Macha 1
Wind Effects on Semisubmersibles and Other Floating Offshore Structures
  E. T. D. Bjerregaard and S. O. Hansen 13
Wind Loads on Offshore Drilling Platforms
  B. J. Vickery, P. Freazy and S. Helliwell 25
Laboratory Simulation of Wind Loads on a Tension Leg Platform
  A. Kareem, P. C. Lu, T. Finnigan and S. L. Liu 43
A Wind/Wave Tunnel Study of Combined Wind and Wave Loads on a TLP
  P. J. Vickery and A. G. Davenport 55
Amplification of Wind Effects on Compliant Platforms
  G. R. Cook, T. Kumarasena and W. Simiu 59

Subject Index 71

Author Index 73

Modeling Wind Loads on Mobile Offshore Structures--
A Summary of Wind Tunnel Results

J. M. Macha*

This paper summarizes the results of several wind tunnel studies of mobile offshore structures conducted at the Environmental Aerodynamics Laboratory of Texas A&M University. The studies included semisubmersible and jack-up drilling units, and a jacket platform during tow. Fundamental aerodynamic phenomena including wind resistance of lattice structures, use of spoilers to eliminate vortex shedding, contribution of lift to overturning moment and effects of the shape and proximity of the sea surface are discussed. Concurrently, various physical modeling techniques to cope with problems associated with model scale and the air-sea interface are addressed. Summary wind tunnel data are presented for several model configurations.

Introduction

Since the late 1970's, the wind tunnel has become increasingly important as a means to determine the wind loads on mobile offshore structures. Previously, wind load predictions were based on simple empirical methods that assign a shape-dependent drag coefficient to each component of the structure. These procedures were originally specified by classification agencies such as the American Bureau of Shipping (ABS) and Det norske Veritas (DnV) for the design of ships, and later adapted to other mobile structures. Early on, it was recognized that such generalized procedures must lead to overly conservative wind loads for some structures, in order to yield safe designs in every case. The first quantitative realization of this situation was provided in a paper by Bjerregaard and Velshou (1978) which compared wind tunnel-based wind loads for a semisubmersible with calculations recommended by the ABS. At moderate to large angles of heel, the overturning moments predicted by the ABS method were considerably greater than moments measured in the wind tunnel. Since the wind overturning moment directly affects both the stability limit and the deck load carrying capacity of a semisubmersible, the pessimistic result of the calculation method would negatively impact vessel cost/performance. In the wake of the Bjerregaard and Velshou study, there has been an increasing dependence on wind tunnel tests to obtain accurate wind load predictions for various types of structures used in offshore exploration and production.

In the context of this paper, the terminology "mobile offshore structure" includes semisubmersibles and jack-up units, but also fixed structures like jacket platforms while under tow. For such structures, wind loads are important for design and operating decisions concerning

*Associate Professor, Aerospace Engineering Department, Texas A&M University, College Station, TX 77843

towing resistance, anchoring and dynamic positioning, deck load carrying capacity, and intact and damaged stability. In many instances, the faults found with present calculation methods stem from their inability to accurately account for three complicated aerodynamic effects: 1) shielding and interference among structural components and the air-sea interface; 2) substantial lift forces; and 3) coupling between the unsteady wind and compliant parts of the structure. The third aspect listed above is still in its infancy, both in terms of physical modeling in the wind tunnel and application to the design process. Other authors in this symposium, and in particular Davenport (Davenport and Hambly, 1984), are making pioneering contributions in this area. As for the first two items, the wind tunnel has the potential to provide quantitative design wind loads for even the most complicated structural layout.

During the past five years, the Environmental Aerodynamics Laboratory at Texas A&M University has conducted dozens of wind tunnel studies of offshore structures. The test articles have ranged from near full-scale models of isolated structural components to reduced-scale models of complete units. Much has been gained in exposing specific weaknesses in the ability of empirical calculation procedures to accurately predict wind loads. At the same time, experimental procedures are still evolving in an effort to achieve the highest possible degree of similitude between model and prototype. Challenges to the experimentalist are manifest in two general categories: 1) those due to model scale effects; and 2) those related to the air-sea interface. The aim of this paper is to present a synergetic summary of the activities of this one laboratory to deal with these two areas to obtain reliable predictions of wind loads in the offshore environment.

Scale Effect Phenomena

The flow around a body depends on both inertial and viscous effects, the ratio of which defines the dimensionless Reynolds number $Re = VL/\nu$, where V is the flow velocity, L a characteristic length of the body, and ν the kinematic viscosity of the fluid. If the Re of the model in the wind tunnel is identical to that of the full-scale structure at the design wind speed, then all of the aerodynamic properties are exactly simulated. However, the large size of offshore structures prevents the realization of this perfect similitude in available wind tunnels. Models of complete units are typically 100 to 200 times smaller than the prototype. Even for tests of isolated components, the model scale rarely exceeds 1:4.

Fortunately, experience has shown that under certain circumstances the important features of the flow patterns that determine the wind load can be adequately replicated at less than the prototype Re. This situation occurs naturally for model geometries with sharp edges that fix the location of flow separation, independent of wind speed or scale. Consequently, wind loads on most deck components and superstructure of semi-submersibles and jack-ups are not susceptible to scale effects.

On the other hand, the separation location on a curved surface, and thus the pressure distribution which determines the wind force, is very dependent on Re. For the particular case of columns and braces of circular cross section, the appropriate length scale is the diameter, and the correlation of separation location and drag with Re is well known. At low values of Re, the flow in the fluid boundary layer adjacent to a circular section is laminar, and separation takes place on the windward face at approximately 80-degrees from the stagnation line. This condition produces a large wake and relatively high drag, and is referred to as a "subcritical" flow pattern. At a sufficiently high value of Re, the boundary layer becomes turbulent before separation takes place. The important feature of a turbulent boundary layer is its ability to proceed farther along the surface before separating. In fact, the separation line moves to the leeward face of the cylinder and the concurrent narrowing of the wake reduces the drag. The decrease in drag occurs over a relatively small increase in Re, referred to as the "critical" range. As Re increases further, the flow pattern becomes "supercritical" and the turbulent separation line migrates upstream, once again increasing the wake width and drag. However, the supercritical drag is still less than that for subcritical flow by nearly a factor of two.

The full-scale Reynolds numbers for structural components of circular cross section are typically well into the supercritical range, with respect to either wind or current. By comparison, values of Re for a wind tunnel model of a complete offshore structure are likely to be subcritical or just approaching the critical range. In this situation, the more subtle details of the transition from subcritical to supercritical flow become important. The transition process is strongly influenced by both surface roughness and the level of turbulence in the approaching flow. Qualitatively, an increase in either of these properties will decrease the Re at which transition occurs.

For a particular wind tunnel test, the relative surface roughness may vary by an order of magnitude depending on model construction methods, and the turbulence level may vary by two orders of magnitude depending on the type of upstream tailoring of the wind that is used. A detailed empirical analysis of the combined effects of roughness and turbulence will usually determine whether the naturally occuring flow patterns for various model components will be subcritical or supercritical. For those components found to be subcritical, it may be possible to improve flow similitude by one of two modeling techniques.

If the value of Re is sufficiently high, transition to supercritical flow can be promoted by judiciously roughening the surface of the component. Sand roughness in a standard grit size over an adhesive coating is commonly used, although a sometimes more convenient and more permanent method is to machine a prescribed surface finish. The required degree of roughness is a function of both Re and the turbulence level, and can be determined experimentally by trial tests. The equivalent grit size should not exceed one percent of the component's diameter, to prevent a significant increase in friction drag on the attached flow part of the surface. Consequently, for a given level of wind turbulence, there is a lower limit on the range of Re where artificial roughness is effective. In the highly turbulent flow of a simulated atmospheric boundary layer, this lower limit on Re is of the order of 1×10^4.

For cylindrical model components with Re less than 1×10^4, the subcritical character of the flow cannot be altered. However, exaggeration of the wind load on those components can be eliminated by reducing

their diameters to approximately one-half the model-scale diameter.
This distorted scaling may be applied to columns and braces with length-to-diameter ratios of five or greater. For smaller ratios, the non-linear dependence of the drag coefficient on the ratio should be considered. In the case of very dense truss-type structures, the resulting decrease in solidity may also influence the overall drag.

The examples discussed in the following subsections outline the primary aspects of current test techniques to deal with scale effects, as well as present typical wind loading data.

Array of circular columns
The objective of the wind tunnel test of the model shown in Fig. 1 was to investigate the effects of surface roughness, wind turbulence and Reynolds number on the drag of an arrangement of columns representative of a semisubmersible. The columns were treated with five different sand roughnesses, covering a range in the ratio of grit size to column diameter, k/d, of 3.3×10^{-3} to 2.9×10^{-2}. The tests were conducted in uniform approach flows of two different turbulence intensities of 0.003 and 0.040. The 0.003 intensity is of interest because the below-water parts of semisubmersibles are commonly tested in such low turbulence flows. The higher turbulence intensity of 0.040 was produced by a grid installed upstream from the model. Turbulence intensities in actual and simulated atmospheric boundary layers are still higher by a factor of four.

Test results for a bow wind and a quartering wind are shown in Fig. 2. The curves for the bow wind do not display the subcritical/supercritical behavior typical of isolated cylinders. The relatively small distance between columns along the wind direction gives the appearance to the flow of a single elongated body. On the other hand, the curves for the quartering wind exhibit the classic sensitivity to Reynolds number. With a smooth surface, the decrease in drag coefficient with increasing Re is indicative of the approach to the critical flow pattern. It is also apparent that the increase in stream turbulence promotes the onset of the transition process. With the addition of the smallest artificial roughness, the drag coefficient reaches a minimum value and then begins to increase in the supercritical region. Further increases in the size of the roughness move the critical region to still lower values of Re.

Figure 2 also shows the strong influence that the sheltering of downwind columns has on the overall drag. The drag for the bow wind is nearly a factor of two smaller than for the quartering wind. The spacing of column centers is 3.13 diameters longitudinally and 8.13 diameters transversely.

Jacket platform during tow
Figure 3 shows a 1:200 model of a jacket platform in the tow configuration. For this structure, the full-scale Reynolds numbers for the tubular members are in the supercritical range of 4×10^5 to 3×10^6. At the model scale, the Reynolds numbers are approximately two orders of magnitude lower. Even considering the turbulence in the simulated wind profile and wake turbulence created by the members themselves, the flow around the model will still be of the subcritical type. Furthermore, the Reynolds numbers of all but the largest members are too low for artificial surface roughness to be effective in promoting transition.

Since the tubular members dominate the wind exposure for this type of structure, a strictly geometrically-scaled model would grossly overpredict the wind load. To obtain more accurate test results, the diameters of all members were reduced further by a prescribed factor designed to compensate for the difference between subcritical and supercritical values of the drag. If the mutual interference among members can be ignored, the factor would simply be the ratio of supercritical to subcritical two-dimensional cylinder drag coefficients. This approach is supported by experimental data for truss structures that show an approximately linear relation exists between overall drag and solidity, for solidities typical of the jacket platform.

On this basis, the wind tunnel model was built at a scale of 1:200, but with tube diameters further reduced by the factor 0.7/1.2. The presumed value of 0.7 for the supercritical drag coefficient of a cylinder was intended to retain some conservatism.

Truss-type legs
The accurate prediction of wind loads on the truss-type legs of jack-up mobile drilling units is important for both structural design and the analysis of vessel stability. In the afloat condition, up to 90 percent of all overturning moments are the result of wind forces on the legs (Smith, et al, 1983). Empirical calculation methods are either of the "building block" type in which the drag of the leg is calculated as the summation of contributions from each of the components, or they consider the leg as a whole with the drag determined on the basis of its solidity. In either method, there is a heavy dependence on the results of wind tunnel experiments.

Since the cross braces between the corner chords are typically circular in cross section, wind tunnel tests of model legs must contend with scale effects. Figures 4 and 5 show models of a particular leg design at two scales. The 5-bay model in Fig. 4 has a scale of 1:12, and was installed vertically on a turntable to investigate wind loads as a function of wind direction. The test section blockage presented by this size model is near the maximum that can be corrected for with confidence. To accommodate a larger scale, and hence increase the test Reynolds number, the approximately 1:5 half-plane model shown in Fig. 5 was also tested. This model uses the tunnel floor as the dividing stream surface along the plane of symmetry. Only one of the four bays was metric, in order to reduce the side-wall effects and the load on the balance. The maximum Reynolds number achieved with the half plane model was still a factor of four less than the prototype design value.

Test results for a configuration similar to that shown in Figs. 4 and 5, but without exposed gear racks on the circular corner chords are presented in Fig. 6. The curve for the 1:5 model exhibits two critical Reynolds number ranges--one for the large diameter corner chords and another for the smaller-diameter braces. When narrow strips of roughness were applied along the length of the members at appropriate circumferential locations, supercritical flow was obtained for the braces independent of Re. The curve for the smooth-surfaced 1:12 model follows the trend of the larger model, but it is only fortuitous that the drag coefficient at its highest Reynolds number is reasonably close to the ultimate supercritical value.

Alleviation of vortex shedding

An alternate leg design for a jack-up unit is an internally stiffened circular column, as dipicted by the model shown in Fig. 7. While the steady drag characteristics are more predictable than for a truss-type leg, the single-column design is susceptible to vortex-induced oscillatory loads. One of the most effective aerodynamic devices used to alleviate vortex shedding is a helical strake, an example of which is seen on the model.

Wind tunnel tests at a model scale of 1:120 validate the effectiveness of the strakes, as shown by the comparative wake spectra in Fig. 8. The strakes increase the steady drag of the entire unit in the elevated mode by approximately four percent. Additional information on the design and use of helical strakes is given a report by Scruton (1963).

Air-Sea Interface

The main feature that distinguishes an analysis of wind loads on a marine structure from the broader parent field of industrial aerodynamics is the free-surface boundary at the air-sea interface. The design wind condition for most analyses is synonymous with a fully-arisen sea, with wave heights comparable to dimensions of the structure. A major contributor to the dynamic signature of the wind field are the perturbations induced by the undulating sea surface. The mean wind profile is also affected, since it is determined by the aerodynamic "roughness" of the sea surface. And there is still speculation about the significance of spray droplet impact on the total environmental loading for marine structures. The monograph by Kitaigorodskii (1970) is a comprehensive treatise on the physics of air-sea interaction.

In the wind tunnel it is not possible to simulate the dynamic coupling between the wind and the boundary that represents the surface. Instead, models are tested over a flat boundary, in a turbulent wind with a prescribed mean profile. But even with this simplification, there are strong aerodynamic interference effects caused by the proximity of the model to the surface. These effects are not predictable with present calculation methods, thus emphasizing the requirement for wind tunnel tests.

The following subsections give examples of the influence of the air-sea interface on wind loads and its treatment in the wind tunnel.

Simulation of the air-sea interface

The conventional method of testing marine structures is to allow those parts of the structure that break the water surface to protrude through cutouts in the floor of the wind tunnel test section. An example of this type of set-up is shown in Fig. 7, where the model of the jack-up unit is in the elevated, drilling configuration. However, to obtain wind load data pertinent to the determination of a floating vessel's stability, the model must be tested at various heel angles. For each orientation, considerable care and time must be given to maintaining minimum clearance between the model and floor cutouts. Since most wind tunnels base their charges on occupancy time, it may become prohibitively expensive to test all of the combinations of draft, wind heading and heel desired. In an attempt to alleviate this problem, an alternate technique using a liquid boundary was initiated by this laboratory several years ago (Ribbe and Brusse, 1981). An example of the method applied to a jack-up unit in the afloat mode is shown in Fig. 9.

Use of the liquid boundary is a testing convenience, and there is no presumption that the resulting wavy surface simulates the air-sea interface. Two separate studies (Troesch, et al, 1983; Macha and Reid, 1984) compared liquid and solid boundary wind tunnel data for semisubmersibles. Those investigations showed that acceptable results can be obtained with a liquid boundary, if surface deformation is kept small by reducing the wind speed. Care must also be taken that the surface deformation does not cause differential hydrostatic loads on submerged parts of the model or its support system. This author is casually aware of one Japanese laboratory that uses a water boundary with a high-viscosity surface film to suppress the formation of waves.

Semisubmersible overturning moment
As mentioned in the introduction, it was the measurement of semisubmersible overturning moments (Bjerregaard and Velshou, 1978) that first pointed out the inadequacy of calculation methods. The comprehensive semisubmersible study by Macha and Reid (1984) confirmed that lift forces can actually produce a restoring moment at large heel angles. Specifically, the center of action of the differential pressure force acting on the deck moves toward the downwind edge of the deck as the clearance with the boundary surface decreases. This aerodynamic interference effect is strongly dependent on vessel geometry, including the ratio of deck dimensions, number of columns, solidity of below-deck braces, and placement of superstructure. The helideck can be a very effective producer of lift and moment as the vessel heels, depending on its location relative to the wind direction.

Figure 10 shows an example of the variability in overturning moment as a function of heel angle that can occur between different semisubmersible designs.

Current loads on semisubmersible hulls
While experimental determination of the hydrodynamic loads on submerged parts of a structure have traditionally been carried out in towing tanks, the wind tunnel has the potential to provide the same data more economically. However, the aerodynamic interference between the model and the boundary is even more pronounced because the vessel's draft is typically much smaller than the lateral dimensions of its structure.

In the particular case of twin-hulled semisubmersibles, lift forces can have a large and unpredictable influence on the moment, and thus on the location of the center of reaction required for stability calculations. As the vessel heels, both the magnitude and distribution of lift produced by the slender hulls are extremely sensitive to the angle of incidence relative to the flow. At heel angles where the ends of the hulls break the water surface, the uncertainty in the measured loads is high whether a liquid or a solid boundary is used.

Static wavy boundary
In a high sea state, perturbations to the magnitude and direction of the wind occur as large waves and swell move past a structure. These

cyclic variations in the wind field can be expected to lead to relatively low frequency, coherent fluctuations in the wind load. The semisubmersible study cited previously (Macha and Reid, 1984) included an idealized simulation of this effect, in which steady wind loads were measured for different model locations along a fixed wavy boundary. The test set-up is illustrated in Fig. 11.

Near the wavy boundary, the dynamic pressure of the wind at a crest was approximately 50 percent greater than in a trough. A corresponding variance of the same magnitude was observed in the overturning moment.

While the sinusoidal boundary used in the study was a grossly simplified representation of an irregular sea surface, it can be argued that the magnitudes of the wind and load perturbations depend primarily on the ratio of wave height to wave length, and secondarily on the actual wave shape. The test results clearly suggest that wave-induced wind fluctuations should be included in any dynamic loads analysis.

Final Remarks

Wind tunnel tests have become commonplace as a means to determine the wind loads on mobile offshore structures. In 1982, the Norwegian Maritime Directorate published new regulations requiring that wind overturning moments for mobile drilling units operating in Norwegian territorial waters be derived exclusively from wind tunnel data, and the U.S. Coast Guard is considering similar action. Established empirical calculation methods are not being challenged on their ability to ensure a safe design, but rather on their inherent exaggerated conservatism that may impact the economics of offshore exploration and production.

While the wind tunnel has demonstrated its potential to predict wind loads more accurately when complicated aerodynamic effects are involved, test procedures are still relatively new and evolving. There is still much to be learned in dealing with scale effects and the treatment of the air-sea interface. Beyond the immediate application of test results to specific structures, there is a need to 1) gain a better fundamental understanding of wind loads in the marine environment; 2) use the results to guide the improvement of the less-costly empirical calculation methods; and 3) refine and standardize model test procedures.

References

Bjerregaard, E.T.D. and Velshou, S. (1978). Wind Overturning Effect on a Semisubmersible, Paper No. 3063, Offshore Technology Conference, Houston, Texas.

Davenport, A.G. and Hambly, E.C. (1984). Turbulent Wind Loading and Dynamic Response of Jackup Platform, Paper No. 4824, Offshore Technology Conference, Houston, Texas.

Kitaigorodskii, S.A. (1970). The Physics of Air-Sea Interaction, English Translation, Israel Program for Scientific Translation, Jerusalem, 1973. (Avail. from N.T.I.S.).

Macha, J.M. and Reid, D.F. (1984). Semisubmersible Wind Loads and Wind Effects, <u>SNAME Transactions</u>, Vol. 92, pp.85-124.

Ribbe, J.H. and Brusse, J.C. (1981). Simulation of the Air-Water Interface for Testing of Floating Structures, <u>Proceedings</u>: 4th U.S. National Conference on Wind Engineering Research, Vol. 1, Seattle, Washington, July, pp.193-195.

Scruton, C. (1963). Note on a Device for the Suppression of the Vortex-Excited Oscillations of Flexible Structures of Circular or Near Circular Section, NPL Report No. 1012, National Physical Laboratory, Teddington, U.K.

Smith, N.P., Lorenz, D.B., Wendenburg, C.A. and Laird, J.S. (1983). A Study of Drag Coefficients for Truss Legs on Self-Elevating Mobile Offshore Drilling Units, <u>SNAME Transactions</u>, Vol. 91, pp. 257-273.

Troesch, A.W., VanGunst, R.W. and Lee, S. (1983). Wind Loads on a 1:115 Model of a Semisubmersible, <u>Marine Technology</u>, Vol. 20, No. 3, July, pp.283-289.

Figure 1. Idealized 8-column semisubmersible model.

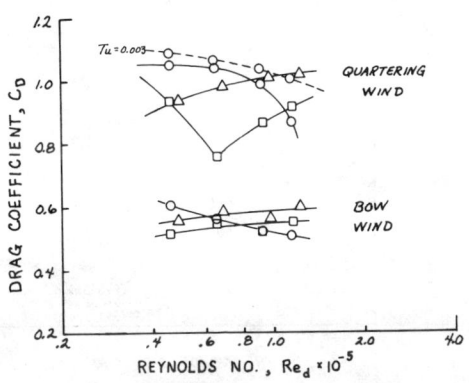

Figure 2. Drag coefficient as function of Reynolds number and surface roughness for the 8-column model.

○ - smooth
□ - $k/d = 3.3 \times 10^{-3}$
△ - $k/d = 2.9 \times 10^{-2}$

Turbulence intensity is 0.04 unless noted.

WIND EFFECTS ON OFFSHORE STRUCTURES

Figure 3. 1:200 model of a jacket platform in tow configuration.

Figure 4. 1:12 model of a truss-type jack-up unit leg.

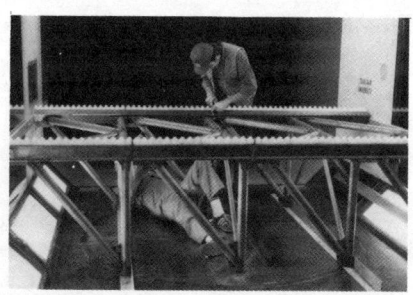

Figure 5. 1:5 half-plane model of the same leg as in Fig. 4.

Figure 6. Drag coefficient of a jack-up leg similar to that shown in Figs. 4 & 5.

○ - 1:12 model
△ - 1:5 half-plane model
□ - 1:5 half-plane model with boundary layer trips on the brace members

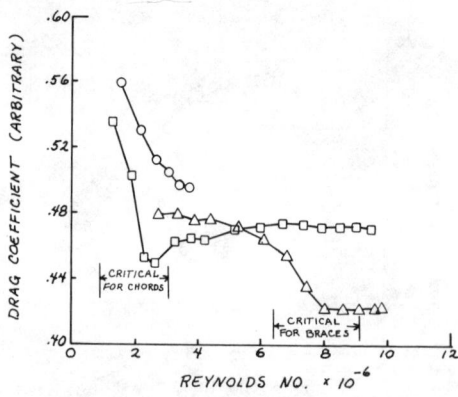

Figure 7. 1:120 model of a jack-up unit with column legs and helical strakes to suppress vortex shedding.

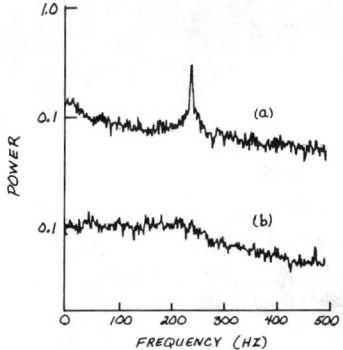

Figure 8. Frequency spectra of pressure fluctuations in the wake of a 1:120 model of a column leg.

(a) without a helical strake
(b) with a helical strake

Figure 9. Jack-up model in the afloat mode with a liquid boundary. Heel angle is 15 degrees.

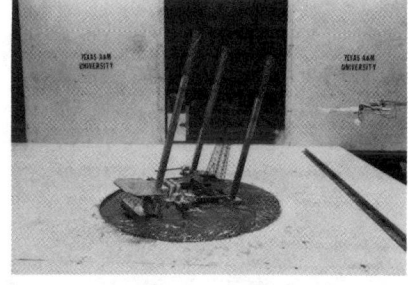

12   WIND EFFECTS ON OFFSHORE STRUCTURES

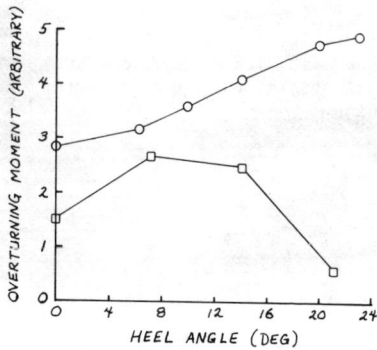

Figure 10. Overturning moment as a function of heel angle for two different semisubmersible designs in a beam wind.

○ - 6-column design
□ - 8-column design

Figure 11. 1:192 model of a semisubmersible over a fixed wavy boundary.

WIND EFFECTS ON SEMISUBMERSIBLES
AND OTHER FLOATING OFFSHORE STRUCTURES.

E.T.D. Bjerregaard * and S.O. Hansen **

ABSTRACT

The paper describes a standardized wind-tunnel test method which is applied for measurements of wind forces and moments on models of semisubmersibles.

By this method, which is in compliance with the requirements of the Norwegian Maritime Directorate, tests are carried out both with the above-water part and the underwater part of the vessel. The main purpose of the tests with the semisubmersible is to obtain data for stability calculations, and hence for determination of the allowable deck load capacity. The vessel is therefore tilted at various angles of inclination and the mean forces and moments are measured for each angle of attack with a six component balance.

Typical examples of results are presented and comparison is made with the results of empirical calculation methods.

Dynamic wind effects on ships are also treated in the paper. The dynamic wind-tunnel test data are typically used as input to a mathematical simulator or a physical wind force generator. The behaviour of the vessel in waves, current and wind can then be studied subsequently either on the simulator or in a model basin.

The nature of the correlation of the atmospheric turbulence is of importance for the wind-induced fluctuating response of the structure. This is illustrated by theoretical calculations of the wind forces on horizontally line-like structures. The calculations are in a reasonable agreement with wind-tunnel measurements carried out in the boundary-layer wind tunnel of the Institute.

INTRODUCTION

Designers and operators of floating constructions need to know as accurately as possible the effects of the harsh environment their constructions are up against.

---

* Head, Wind Engineering Dept., Danish Maritime Institute,
  Hjortekærsvej 99, DK-2800 Lyngby, Denmark
** R & D Manager, Wind Engineering Dept., Danish Maritime Institute.

Wind, wave and current are all determining factors both when considering the structural safety and the operational safety of a floating rig or a ship regardless of whether the vessel is kept in place or moved.

A detailed analysis of the behaviour of a new design most often involves model tests in wave basins and wind tunnels where the extreme weather conditions can be simulated properly. The measured model-test data are subsequently used for prediction of the operational limitations of the design for example in relation to station keeping ability, motion characteristics, surviveability, course-keeping ability, etc. Computer simulations using input data from model tests are important tools in this connection.

This paper is confined to a description of examples of how wind effects on semisubmersibles and ships can be determined from model tests in wind tunnels. The principles applied when simulating dynamic wind forces in a mathematical simulator or in a physical wind force generator are mentioned.

Wind forces and moments on semisubmersibles are of primary concern when establishing the permissible deck load carrying capacity and for mooring analysis. For ships the wind effects influence the course-keeping ability and the manoeuvrability which is of great concern especially when the ship is operating in confined areas for instance in narrow harbours. Under such conditions the fluctuating wind forces are also important.

SEMISUBMERSIBLES

Wind forces and particularly wind overturning moments play a very important role in connection with the determination of the stability requirements of semisubmersibles. According to the basic stability criteria the area under the curve for the hydrostatic righting moment versus inclination angle should be at least 1.3 times the area under the corresponding wind moment curve (Fig. 1). This implies that an

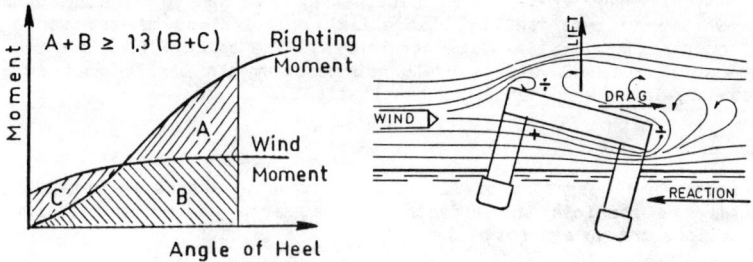

Figure 1. Stability Criteria.　　　　Figure 2. Sketch of the Wind Flow over a Platform.

increase in the wind moment will lead to a decrease in the height of the centre of gravity and hence a reduction in the permissible pay load and vice versa. (Bjerregaard and Velschou, 1978).

Both the authorities and the operators are interested in a determination of the wind moment which is as correct as possible.

In the various codes of practice for calculation of wind effects it is a common method to prescribe drag coefficients in non-dimensional form for different geometrical shapes. The wind force at a given wind velocity is then proportional to the wind speed squared, the shape coefficient and the projected area. The wind moment is the product of the wind force and the distance between the centre of action of the wind force and the centre of reaction. This method does not include lift forces. However, due to the large extension of semisubmersibles in the wind direction, the effect of lift forces on the overturning moment can be of vital importance.

The only way of obtaining exact knowledge of the sizes of the lift and drag forces and the resulting moments is by carrying out windtunnel tests with a carefully scaled model of the semisubmersible.

The Norwegian Maritime Directorate (NMD) requests that all floating rigs under NMD's authority are tested in model scale in the wind tunnel in connection with the assessment of the stability requirements.

In order to provide the necessary data tests are carried out at various angles of inclination around the critical axes for various draughts both for the above-water part and the underwater part of the rig.

Wind Overturning Moment

The origin of the wind forces acting on a semisubmersible can be illustrated via a simple sketch (Fig. 2). When the incoming wind flow passes the semisubmersible a distorsion of the streamlines will be introduced. This will create suction on some areas and pressure on others. Due to the complexity of the rig's geometry and the turbulent structure of the wind it is not possible to calculate accurately the pressure distribution and the resulting horizontal and vertical forces and moments.

In the wind tunnel, these forces and moments are measured directly on the scale model of the rig suspended to a six-component force and moment balance.

The total wind overturning moment equals the wind force, D, in the wind direction multiplied by the distance between the centre of application of the wind force and the centre of reaction of the hydrodynamic forces (Fig. 3).

The wind force and its "centre" is found from tests with the above-water part of the model exposed to a simulated natural ocean wind. The centre of reaction is obtained from tests with the underwater part in a uniform flow simulating a free floating platform. Also for the underwater part it is very difficult or impossible to determine the centre of reaction from empirical calculations. The same complexity of the flow is encountered as for the above-water part, and the corresponding pressure distribution results in both horizontal and vertical force components. The equivalent reaction centre is found from the model test results by dividing the measured moment by the measured drag force.

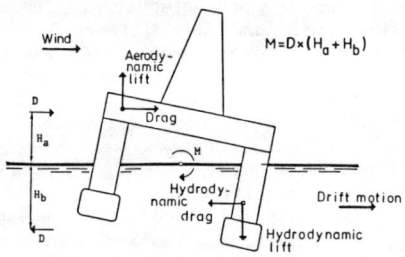

Figure 3. Wind Force and Moment Arm for Calculation of Overturning Moment.

It is important to note that the moment contains the components of the drag as well as the lift forces. Hence, when the moment is divided by the drag force, the equivalent drag centre often turns out to be far away from the centroid of the body. In many empirical calculation methods, the centroid is used as the centre of reaction. Our measurements have shown that, in many cases, this leads to erroneous results.

The forces and moments on the underwater model could just as well be measured in the towing tank as in the wind tunnel. However, it is often more convenient and more economical to perform the tests in the wind tunnel, especially when many different flow directions, draughts and angles of list have to be included in a test programme.

In a previous paper (Bjerregaard and Sørensen, 1982) it was demonstrated from comparative tests in water tank and wind tunnel that the wind-tunnel test technique is a reasonable approach for the determination of the underwater reaction centre in cases where wave-making resistance can be neglected.

Normal Test Procedure

The test procedure most often used is the one required by the Norwegian Maritime Directorate in connection with stability calculations. At least three different loading conditions are investigated.

For each draught, tests are initially carried out with the rig on even keel. Flow directions are covered from 0 to 360 degs. at increments of 10 degs. On the basis of the measured moments of the above-water and underwater part, the wind direction giving the largest overturning wind moment is determined.

An example of a typical result for a drilling rig (scale 1:250) is shown on Fig. 4. The maximum moment is found to be about 315 degs., which corresponds to bow quartering wind from starboard side. Afterwards the model is inclined around an axis perpendicular to the most critical wind direction, and for a number of angles of inclination, f.inst. $5°$, $15°$ and $25°$, measurements are made of wind moments and reaction moments. These tests are carried out for wind directions in the sector of +/- 30 degs. around the critical wind direction for the above-water part and the corresponding opposite directions for the underwater parts. After completion of the same test programme for all three draughts the stability of the rig is evaluated.

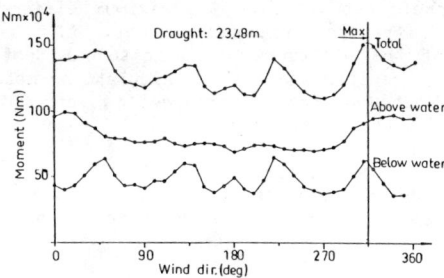

Figure 4. Overturning Moment as Function of Wind Direction. Even Keel.

Wind-Tunnel Tests versus Calculations

During the last 7-8 years DMI has carried out wind-tunnel tests with more than 40 different semisubmersible models for various clients. For the majority of the platforms the above-mentioned test procedure was applied.

There has for a number of the platforms been an opportunity to compare the results from the wind tunnel tests with the results from empirical calculations, MODU-code or ABS.

The comparison has shown the following general trends:

1. Calculated wind drag forces are normally heigher than the wind forces obtained from wind-tunnel tests. This is mainly due to shielding effects and interaction effects which cannot be correctly accounted for in the calculations.

2. The distance from the water line to the centre of action of the wind force (H_a in Fig. 3) is heighly dependent on the

lift forces especially for the inclined conditions. Since lift is not taken into consideration in the calculation method this distance will normally differ considerably. For quartering wind and moderate angles of inclination the lift tends to increase the distance, H_a, whereas for beam wind the lift often reduces H_a for inclined conditions.

3. The distance from the water line to the centre of reaction, (H_b in Fig. 3), shows a similar dependency of the lift as for the above-water part. H_b measured in the wind tunnel (or tank) is normally larger than the calculated H_b for upright condition and small angles of inclination. For larger inclination angles and particularly for the beam wind condition the opposite results are most often found.

4. For about 80% of the compared platforms it was found that for the survival and operational condition the area under the wind moment curve was larger for the calculation method than for the wind-tunnel test method. The maximum difference in area was abt. 25%. (Average difference was abt. 6%).
For a few of the platforms the wind-tunnel tests gave more conservative results than the calculation methods. The largest difference we have experienced is abt. 20% lower wind moment area for the calculations.

As an example the main results from a platform are presented on Figures 5 to 7. The semisubmersible is a drilling rig with two pontoons. Details about the rig cannot be revealed as the data are proprietary.

Measurements were made both for quartering wind and for beam wind. Figure 5 shows that the correlation between the calculated wind force and measured wind force is better for the beam wind than for the quartering wind condition.

Figures 6 and 7 show the total wind overturning moment, M_{total}, and how it originates from the moment on the above-water part, M_a, and the moment on the underwater part, M_b. Figure 6 reveals that although the calculation method gives a larger moment on the above-water part than the wind-tunnel test, the total moment up to an inclination angle of abt. 18 degs. is larger for the wind-tunnel tests. This is because the measured moment on the underwater part is larger than the calculated.

For the beam wind condition shown in Figure 7 the calculated results are clearly on the conservative side. There is a distinct difference in the character of the curves between the model test results and the results of calculations.

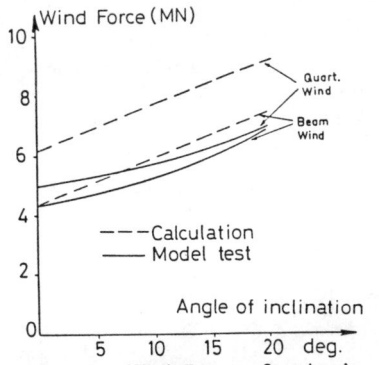

Figures 5-7. Comparison between calculations and wind-tunnel test results for a semisubmersible drilling rig. Draught = 19.5 m. Wind Velocity = 50 m/s.

Figure 5. Wind Force, Quartering Wind and Beam Wind.

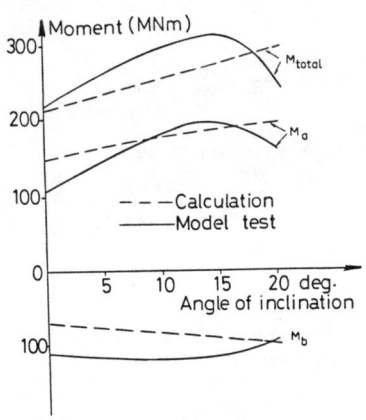

Figure 6. Overturning Moment, Quartering Wind

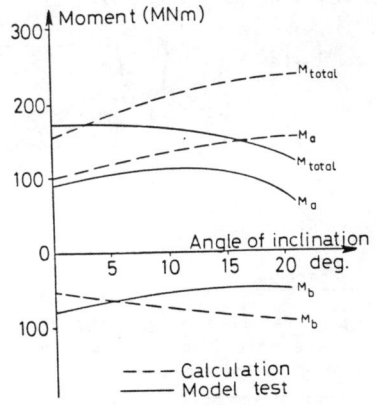

Figure 7. Overturning Moment Beam Wind.

DYNAMIC WIND EFFECTS - THEORETICAL CONSIDERATIONS

The fundamental natural period of large compliant offshore structures is typically of the order of 100 seconds. This indicates a natural frequency close to the region of maximum energy in the spectrum of atmospheric turbulence.

The nature of the correlation of the atmospheric turbulence is of importance for the structural response. This is illustrated below by calculations of the wind forces on horizontally line-like structures. It is a simplified representation of e.g ferries and tankers.

The lateral force and the yawing moment on a ship are considered for the beam wind condition. The theoretical calculations are compared with wind-tunnel measurements carried out in the boundary-layer wind tunnel described by Hansen & Sørensen (1985).

Aerodynamic Admittance Functions

Assuming quasi-steady conditions, the forces on line-like structures to a turbulent wind can be estimated theoretically by methods originally proposed by Davenport in the beginning of the sixties. The methods are outlined in detail by Davenport (1977).

The spectra of the lateral force and the yawing moment can be expressed by:

$$S_Y(f) = (\rho \cdot C_Y \cdot A_s \cdot U)^2 \cdot S_u(f) \cdot |X_Y(f)|^2$$

$$S_N(f) = (\rho \cdot C_Y \cdot A_s \cdot L \cdot U)^2 \cdot S_u(f) \cdot |X_N(f)|^2$$

where $S_u(f)$ is the spectrum for the longitudinal component of atmospheric turbulence and $|X_Y(f)|^2$ and $|X_N(f)|^2$ are the Aerodynamic Admittance Functions given by:

$$|X_Y(f)|^2 = \int_0^1 \int_0^1 \sqrt{Coh_u(x_1,x_2,f)} \, dx_1 \cdot dx_2$$

$$|X_N(f)|^2 = \int_{-\frac{1}{2}}^{\frac{1}{2}} \int_{-\frac{1}{2}}^{\frac{1}{2}} \sqrt{Coh_u(x_1,x_2,f)} \, x_1 \cdot x_2 \cdot dx_1 \cdot dx_2$$

x is a normalized coordinate along the structure. The phase spectrum of the turbulent component u is assumed to be zero.

Root-coherence

The root-coherence for the atmospheric turbulence in a plane normal to the mean wind flow is recommended to have the following form by several researchers

$$\sqrt{Coh_u(x_1,x_2,f)} = \exp(-C \cdot f \cdot d/U)$$

The expression indicates that the root-coherence approaches unity for small frequencies f. However, this is not the case in practice for large separations d, where the wind structure is characterized by a lack of correlation even at low frequencies. Further, the size of the root-coherence at large separations and at low frequencies are quite important in connection with compliant offshore structures.

ESDU 75001 suggests another expression for the root-coherence, which is in reasonable agreement with the simulated wind structure in the boundary-layer wind tunnel of DMI.

The ESDU expression is represented by Bessel functions, but the main features are included by the formula:

$$\sqrt{Coh_u(d,f)} = \exp(-1.14 \cdot h^{1.4})$$

EFFECTS ON SEMISUBMERSIBLES

where ESDU specifies the parameter h as follows:

$$h = 0.747 \cdot \frac{d}{L_1(f)} \sqrt{1 + 70.8 \left(\frac{f \cdot L_1(f)}{U}\right)^2}$$

$$L_1(f) = 2 \cdot L_u \qquad f < 0.008 \text{ Hz}$$

$$L_1(f) = 2 \cdot L_u \cdot \frac{0.04}{f^{2/3}} \qquad f > 0.008 \text{ Hz}$$

$$L_u = 10 \cdot z^{0.38}/z_0^{0.068} \quad (m)$$

It should be emphasized, that the root-coherence does not approach to unity for small frequencies when the separations, d, are large.

The influence of the root-coherence assumption is illustrated on Figs. 8-9 showing the Aerodynamic Admittance Functions for the lateral force and the yawing moment respectively. A typical value of the reduced frequency in connection with large compliant offshore structures could be $f \cdot L/U = 0.01 \cdot 100/50 = 0.02$. The error introduced by using the normal expression is seen to be quite large.

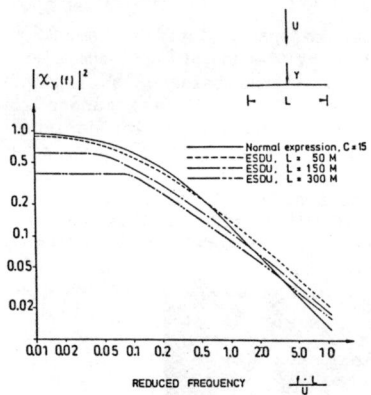

Figure 8. Aerodynamic Admittance Function for the Lateral Force.
$U = 25$ m/s, $L_u = 32.8$ m.

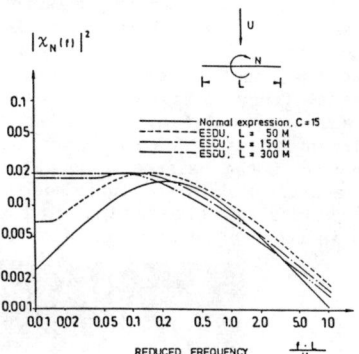

Figure 9. Aerodynamic Admittance Function for the Yawing Moment.
$U = 25$ m/s, $L_u = 32.8$ m.

Root-Mean-Square Loads

The intensity of the lateral force, I_Y, is defined as the ratio of the root-mean-square (rms) value to the mean value. The rms value can be found by integration of the spectrum. This has been done on Fig. 10 showing the intensity values as function of the length of the line-like structure.

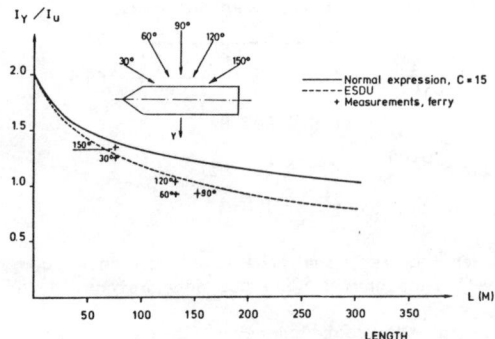

Figure 10. The Intensity of the Lateral Force. $U = 25$ m/s, $L_u = 32.8$ m ESDU spectrum for the turbulent component u.

DYNAMIC WIND EFFECTS - WIND-TUNNEL MEASUREMENTS

Balance Measurements

The mean as well as the dynamic lateral force and yawing moment on a model of a ferry have been measured by means of a 2-component strain-gauge balance. As indicated on Fig. 11 the balance is situated inside the model in order to measure the load in an easy manner at different ferry positions in a harbour area. The balance was designed to ensure a satisfactory characteristic of the system, that is a sufficiently stiff system to obtain a relatively high natural frequency of the balance - model combination. The model of the ferry was manufactured very light to intensify this effect. This enables that undistorted fluctuating wind forces on the model can be determined in the frequency range of interest.

Figure 11. 1:700 Scale Model of a 152 m Ferry with a 2-component Balance situated inside the Model. This model was used in an experiment carried out on behalf of the Danish State Railways.

Measurements

The measured intensities of the lateral force, I_y, on the ferry for the wind directions: $30°$, $60°$, $90°$, $120°$ and $150°$ are shown on Figure 10. The length of the ferry projected on a plane normal to the mean wind flow has been used as x-coordinate. As can be seen there is a good correspondance between measurements carried out in the boundary-layer wind tunnel and the calculated intensity of the lateral force. The somewhat lower intensity in the measurements can be due to the correlation properties in the vertical direction which are not included in the calculations.

SIMULATION OF DYNAMIC WIND EFFECTS

The measured wind loads in the wind tunnel are often used as input to a mathematical simulator or a physical wind force generator. The behaviour of the vessel in waves, current and wind can then be studied subsequently either on the simulator or in a model basin.

In physical modelling the wind loads are simulated by means of the reaction forces from computer controlled air fans placed on the deck of the model.

The wind-tunnel test results with the above-mentioned ferry were used as input data for simulated manoeuvres with a pilot in the loop. The experience from trials with a static wind representation compared with the dynamic wind representation revealed that harbour manoeuvres were significantly more difficult to perform successfully with the dynamic wind present.

CONCLUSION

A preliminary analysis of a number of semisubmersibles has indicated that wind overturning moments obtained from empirical calculations are more conservative than results obtained from wind-tunnel tests for about 80% of the analysed cases. However, for a few platforms the wind-tunnel tests gave larger moments than the calculation method.

It has been demonstrated that lift, shielding and interaction effects, which are not taken into account correctly in the calculation methods, are very important, not only for the determination of wind forces and moments on the above-water part, but also for the assessment of the reaction centre on the underwater part and hence for the total overturning moment used for stability calculations.

Simulated manoeuvring with ships in a mathematical simulator has indicated that the dynamic wind effects are important in harbour areas. Theoretical calculations and measurements in a boundary-layer wind tunnel have shown that the nature of the correlation of the atmospheric turbulence is of importance for the wind-induced fluctuating loads on ships. The correspondance between theory and measurements were found to be reasonable.

NOMENCLATURE

A_s	Projected side area.
C	Constant defining the wind coherence function.
Coh	Coherence function.
C_Y	Non-dimensional lateral force component, $C_Y = \dfrac{Y}{\frac{1}{2}\rho U^2 \cdot A_s}$
d	Separation.
D	Wind drag force.
f	Frequency.
H_a	Moment arm, above-water part.
H_b	Moment arm, underwater part.
I_u	Turbulence intensity at deck height of the ferry (12%).
I_Y	Intensity of the lateral force.
L	Length of the ship.
L_u	Length scale parameter.
M_{total}	Total overturning moment.
M_a	Overturning moment, above-water part.
M_b	Overturning moment, underwater part.
N	Yawing moment.
S	Spectrum
u	Turbulent component in the wind direction.
U	Mean wind velocity.
x	Normalized coordinate along the structure.
Y	Lateral force.
z	Height above sea level.
z_0	Roughness length.
ρ	Air density
$\|\chi\|^2$	Aerodynamic Admittance Function.

REFERENCES

Bjerregaard, E.T.D. and E.G. Sørensen (1982): "Forces and Moments on Underwater Bodies Obtained from Wind-Tunnel Tests". OTC 4438.

Bjerregaard, E.T.D., S. Velschou and J.S. Clinton (1978): "Wind Overturning Effect on a Semisubmersible". OTC 3063.

Davenport, A.G. (1977): "The Prediction of the Response of Structures to Gusty Wind". International Research Seminar on Safety of Structures under Dynamic Loading, Trondheim, Norway, June 23 to July 1, 1977.

Hansen, S.O. and E.G. Sørensen (1985): "A New Boundary-Layer Wind Tunnel at the Danish Maritime Institute". J. Wind. Eng. Ind. Aerodyn., 18 (1985) p. 213-224.

ESDU 75001: "Characteristics of Atmospheric Turbulence near the Ground". Engineering Sciences Data Unit (1975).

WIND LOADS ON OFFSHORE DRILLING PLATFORMS

B.J. Vickery, P. Freathy[*] and S. Helliwell

The University of Western Ontario
London, Ontario, Canada

[*]presently, British Telecom, London, England

1.0 INTRODUCTION

Wind loads on conventional offshore platforms constitute only a small fraction of the design load which is determined primarily by wave action. Compliant structures, such as the guyed tower or tension-leg platform, are designed to minimize wave loading and, as a result, the wind forces play an important and possibly dominant role. The present paper is concerned primarily with wind load studies conducted at the Boundary Layer Wind Tunnel Laboratory of The University of Western Ontario over the past five years. These studies have included a very comprehensive study of the mean and dynamic loads on a particular platform and a parametric study on a somewhat simplified set of models. The results of these studies have been reported in part by Pike and Vickery (1982), Vickery and Pike (1985), Freathy and Vickery (1986) and Freathy (1985). The published results are reviewed and are supplemented by findings derived from more recent studies and results not previously reported.

2.0 MEAN WIND LOADS

2.1 The "Lena" Guyed Tower

The major results of this study are reported by Pike and Vickery in References 1 and 2. The most notable feature of this study was the attention paid to modelling and the effects of model simplification. The original study (1) was conducted using a 1:120 model in which the details extended to accurate modelling of stairways (Fig. 1). Reynolds Number effects on sharp-edged structures do not become significant above about Re = 400 and at a scale of 1:120 and test speeds of the order of 10 m/s this permits the scaling of typical structural elements with full scale dimensions as small as 0.1m. Members of circular cross-section provide greater scaling difficulties but the procedure of undersizing such members at model scale so as to produce equality of $C_d \cdot d$ (C_d = drag coefficient, d = diameter) is an acceptable modelling technique. The values of C_d at both model and prototype scale are well defined and the size reduction also produces a wake width which is well scaled and hence it can be expected that shielding effects are also reasonably modelled. The second phase of the "Lena" study employed a comparatively simple model constructed at a scale of 1:400 (Fig. 2). At this scale the details could not be reproduced but the results obtained in regard to mean loads compared favourably with those obtained at a sale of 1:120. A comparison of the drag force coefficients is shown in Fig. 3 and these results are typical. Variations average a few percent and do not exceed 6% of the maximum drag coefficient.

Fig. 1: 1:120 model at trough of fixed design wave

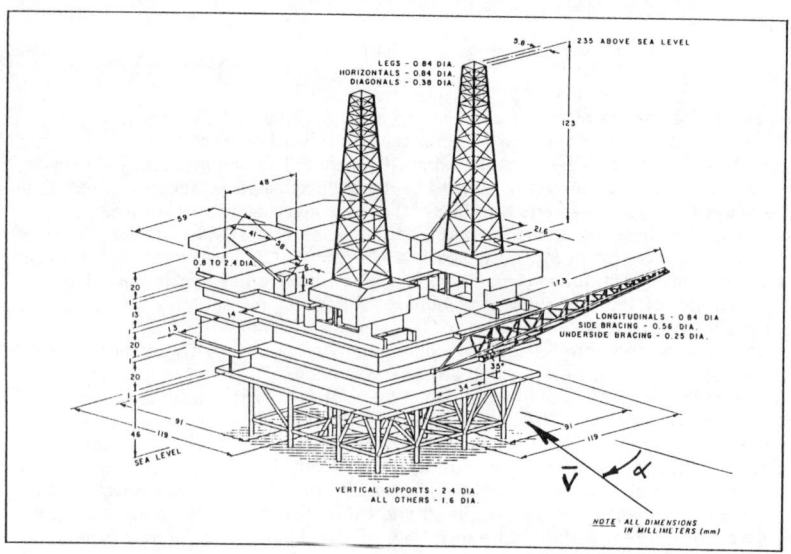

Fig. 2: Principal dimensions of 1:400 model

LOADS ON DRILLING PLATFORMS 27

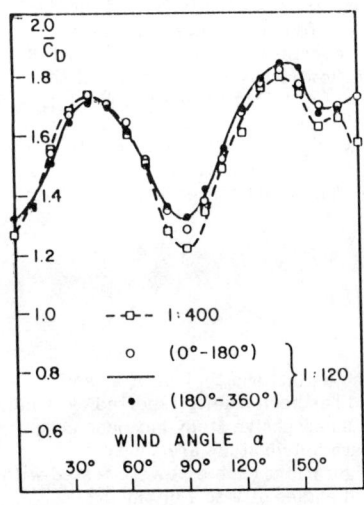

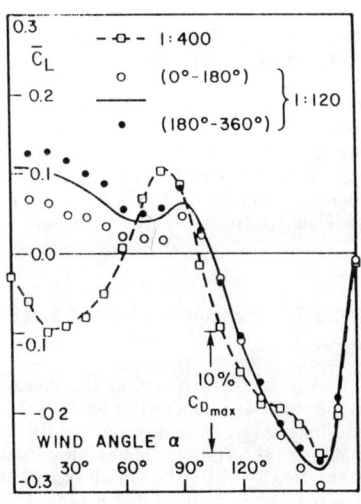

Fig. 3: Mean drag force coefficients

Fig. 4: Mean transverse force coefficients

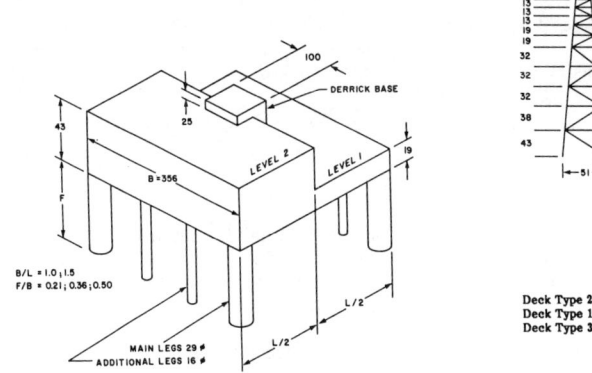

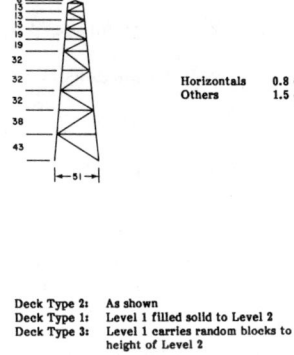

Deck Type 2: As shown
Deck Type 1: Level 1 filled solid to Level 2
Deck Type 3: Level 1 carries random blocks to height of Level 2

Fig. 5: Dimensions of models employed in parametric study

A comparison of the measured forces with those predicted by the methods of the American Petroleum Institute (A.P.I. rules) are reported in (1) and for the maximum drag loads the A.P.I. predictions were found to overestimate the measured values by 36% for load and 18% for overturning moment. The A.P.I. rules do not cover forces normal to the mean flow (transverse forces) but the model studies showed these to be 15% to 20% of the drag force (Fig. 4). Lift forces were also measured and were found to be very significant. At maximum drag the lift forces for the configuration shown in Fig. 1 amounted to 40% of the drag force and with the two derricks removed, as would be the case for a production platform, this ratio attained 80%. These lift forces also contribute significantly to the overturning moment.

2.2 Parametric Study of Mean Loads

A parametric study of the mean loads on a platform typical of a semi-submersible was conducted by Freathy (4) and has been briefly reported by Freathy and Vickery (3). A somewhat similar but not as extensive study has been described by Macha (5). The models employed in the parametric study are shown schematically in Fig. 5. In all, some 54 configurations (Table 1) were tested with six force components being measured for wind angles of 0 to 180° in 15° increments. The complete data set is reported in reference (4) and only selected results and observations are presented here. The force and moment coefficients are defined as;

$$C_F = F/\tfrac{1}{2}\rho V^2 B^2$$
$$C_M = M/\tfrac{1}{2}\rho V^2 B^2$$
V = mean speed at height B
ρ = air density
F = force
M = moment.

The influence of the various parameters on the force and moment coefficients are shown in Figs. 6 to 8. Fig. 6 shows the effect of heel angle on the drag force. Fig. 7 shows the influence of deck type on drag force and moment while the influence of the free-board ratio (F/B) is shown in Fig. 8.

The measured coefficients were compared with those computed in accordance with the following approaches;

Series A: Square Planform

No.	Freeb'd ratio	Heel (deg)	Deck Type	H'deck	Der'k	No of Legs
1	0.50	0	1	no	yes	4
2	0.50	10	1	no	yes	4
3	0.50	20	1	no	yes	4
4	0.36	0	1	no	yes	4
5	0.36	5	1	no	yes	4
6	0.36	10	1	no	yes	4
7	0.36	15	1	no	yes	4
8	0.36	20	1	no	yes	4
9	0.21	0	1	no	yes	4
10	0.21	10	1	no	yes	4
11	0.21	20	1	no	yes	4
12	0.50	0	2	no	yes	4
13	0.50	10	2	no	yes	4
14	0.50	20	2	no	yes	4
15	0.50	0	3	no	yes	4
16	0.50	10	3	no	yes	4
17	0.50	20	3	no	yes	4
18	0.50	10	1	no	yes	8
19	0.50	20	1	no	yes	8
20	0.36	0	1	no	yes	8
21	0.36	10	1	no	yes	8
22	0.36	20	1	no	yes	8
23	0.21	0	1	no	yes	8
24	0.21	10	1	no	no	4
25	0.50	0	1	no	no	4
26	0.50	10	1	no	no	4
27	0.50	20	1	no	no	4
28	0.50	0	1	yes	yes	4
29	0.50	10	1	yes	yes	4
30	0.50	20	1	yes	yes	4

Series B: Rectangular Planform

No.	Freeboard ratio	Inclin'n (deg)	Deck Type	Module Position
1	0.36	0	1	-
2	0.36	10 roll	1	-
3	0.36	20 roll	1	-
4	0.36	10 pitch	1	-
5	0.36	20 pitch	1	-
6	0.21	0	1	-
7	0.21	10 roll	1	-
8	0.21	20 roll	1	-
9	0.21	10 pitch	1	-
10	0.21	20 pitch	1	-
11	0.36	0	3	-
12	0.36	10 roll	3	high
13	0.36	10 roll	3	low
14	0.36	20 roll	3	high
15	0.36	20 roll	3	low
16	0.36	10 pitch	3	-
17	0.36	20 pitch	3	-
18	0.21	0	3	-
19	0.21	10 roll	3	high
20	0.21	10 roll	3	low
21	0.21	20 roll	3	high
22	0.21	20 roll	3	low
23	0.21	10 pitch	3	-
24	0.21	20 pitch	3	-

Table 1: Configurations Tested in Parametric Study

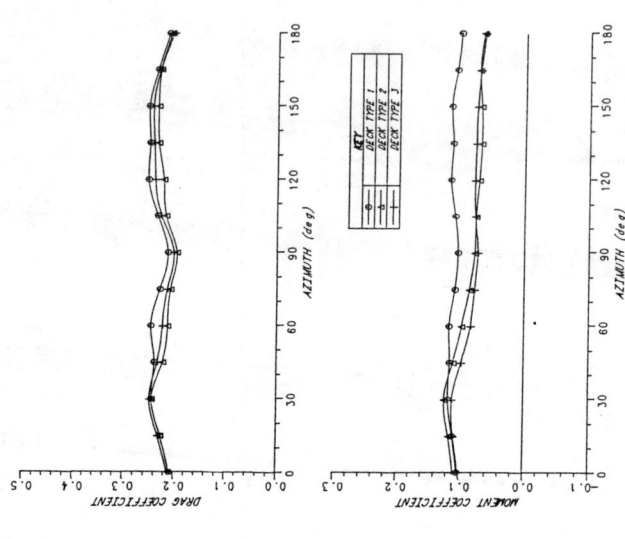

Fig. 7: Influence of deck type on the mean force and moment coefficients (zero heel, square model, F/B=0.5)

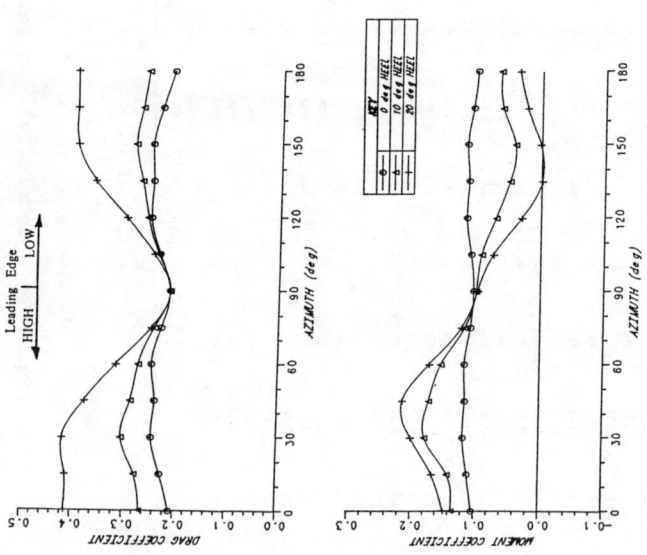

Fig. 6: Influence of heel angle on the mean force and moment coefficients (square model, deck type 1, F/B=0.5)

LOADS ON DRILLING PLATFORMS

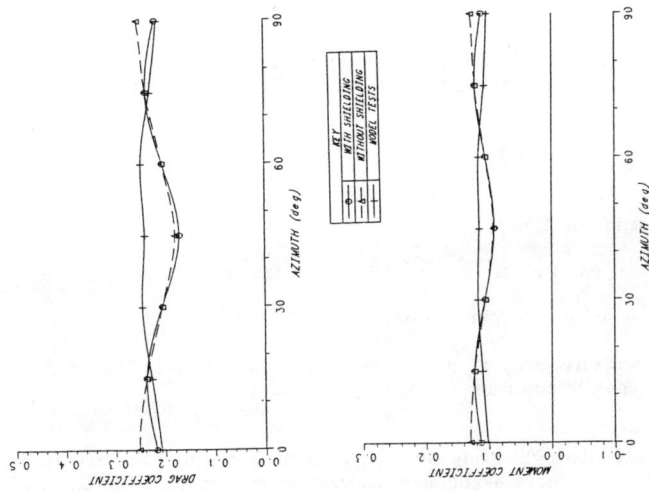

Fig. 9: Comparison of model results with ABS predictions using methods 1 and 2.

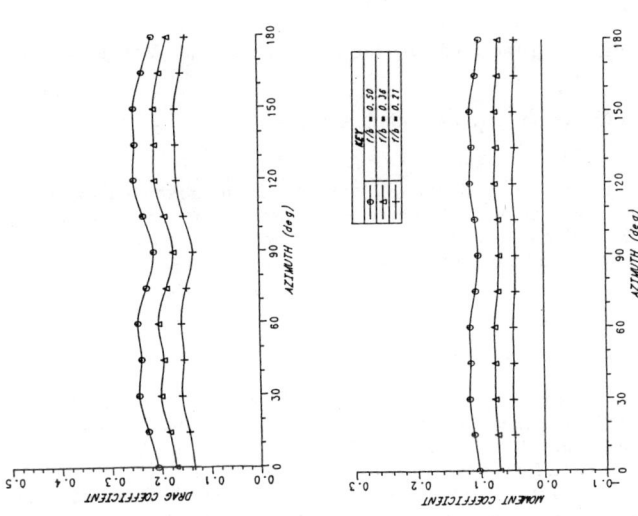

Fig. 8: Influence of freeboard on the mean force and moment coefficients (zero heel, square model, deck type 1)

2.2.1 American Bureau of Shipping (ABS)

Wind loading for fixed platforms is dealt with in paragraph 3.5.2 of reference 6. For the wind normal to flat surfaces the familiar equation is used:

$$F = 0.0623 \, V_y^2 \, C_h \, C_s \, A \qquad \text{(kgf)}$$

C_h is a height coefficient

C_s is a shape coefficient

V_y is the velocity at height y (m/s)

A is the projected area of the member on a plane normal to the direction of the considered force. (m^2)

It further states that: "For any direction of wind approach to the structure, the wind force on flat surfaces should be considered to act normal to the surface." There are three ways of interpreting this statement:

1. Split the velocity vector into orthogonal components normal to the exposed plane surfaces and recombine the calculated forces to give a resultant drag in the wind direction.

2. As above but splitting the wind dynamic pressure $(q = 0.5 \, \rho V^2)$ not velocity.

3. Applying the full velocity to each plane surface.

According to the report by Macha (5), method 2 above is the commonly accepted procedure. Method 3 is considered to be unreasonably conservative and was not studied. Shielding allowances are permitted where the separation between members is less than seven times the diameter of the upwind member.

2.2.2 American Petroleum Institute (API)

The code used for the present study is that for fixed structures. Paragraph 2.3.2 of Reference (7) describes the method for calculating loads. It is essentially the same as that prescribed by the American Bureau of Shipping. The API code requires that "appropriate formulas" be used to calculate the load on flat surfaces which are not perpendicular to the wind and that shielding may be included where the designer judges it to be warranted. The shape coefficients used are similar to those of the ABS rules except for sides of buildings where the recommended coefficient is 50% larger.

LOADS ON DRILLING PLATFORMS 33

2.2.3 British Standards Institute (BSI)

Paragraph 3.3.5.4 of the British Code (8) simply refers the designer to CP3: Chapter V: Part 2: 1972 (9) - the code for building wind loads. CP3 treats the subject of shape coefficients in much more detail that the other codes and makes allowances for the ratio of downwind to crosswind dimension for block-like elements of structure. For cylinders a graph of drag coefficient versus Reynolds Number is given which includes the effect of surface roughness. A shielding allowance is permitted when the downwind distance between members is less than seven diameters. In comparison with the other codes the coefficient for the block-like structures is lower whilst for the cylindrical elements it is higher.

2.2.4 Det Norske Veritas (DnV)

Of the offshore codes which treat the wind loads without reference to a building code, DnV's rules (10) are the most detailed. The shape coefficient for circular cylinders is treated in a similar way to the BSI code (8). There is no specific reference to block-like structures, although paragraph 203 refers to "smooth members of rectangular cross-section". In the absence of other information, these were used for the block elements but they lead to unreasonably high shape coefficients, especially when there is sufficient downwind length to permit reattachment of the flow. A shielding allowance is permitted when the downwind distance between members is less than seven diameters.

The equation to be used for calculating forces acting in the plane of the cross-section of the member is:

$$F = 0.5 \, \rho \, C_S \, V^2 \, b \, l \, \cos^2 \beta$$

C_S is a shape coefficient

V is the flow velocity (m/s)

l is the length of member (m)

b is the cross-sectional dimension perpendicular to the flow (m)

β is the wind angle.

By including $\cos^2 \beta$ the velocity is effectively being split into components.

2.3 Comparison of Computed and Observed Loads

For each of the four codes computations were made in accordance with each of the following four methods;

1. Splitting the velocity vector into orthogonal components and applying them to the two faces of the platform. The forces are then recombined to give the resultant drag and moment in the direction of the wind.

2. As for (1) above but permitting a 50% shielding of the downwind legs when they are directly behind the upstream legs.

3. As for (1) but splitting the dynamic pressure vector ($q = 0.5 \rho V^2$).

4. As for (2) but splitting the dynamic pressure vector.

Comparisons of the measured force coefficients with those calculated are shown in Figs. 9 to 13. The major conclusions arising from these comparison are;

(i) Of the four Codes tested two - the ABS rules (6) and the British Standard (8) - produced overall drag coefficients which were closest to those measured for the simple square model. Of these two only the ABS Rules are specifically intended for use on floating offshore platforms. Two other codes examined - the API Rules (7) and the DNV Rules (10) - gave results which were considerably higher than the measured results.

(ii) The agreement found for drag coefficient values with the four codes was improved if, when considering the effect of azimuth angle, the dynamic pressure vector was split into orthogonal components in order to compute the force on the two faces of the platform. It is recommended that this method should be adopted where the rules for classification permit.

2.4 Modelling the Sea-Air Interface

The studies reported in sections 2.1 and 2.2 were conducted in a boundary layer wind tunnel with a solid floor. The roughness was chosen to produce a profile consistent with flow over the sea but the true interface was not modelled. The 1:120 model of the "Lena" tower was tested with a stationary wave profile as shown in Fig. 1 and although this is far from realistic the variations of load and moment with position on the wave give an indication of the sensitivity of wind loading to wave position. In this case the wave had a length equal to six times the breadth of model and height equal to 60% of the breadth or about 120% of the freeboard. The changes in drag load and overturning moment with position are shown in Table 2.

LOADS ON DRILLING PLATFORMS 35

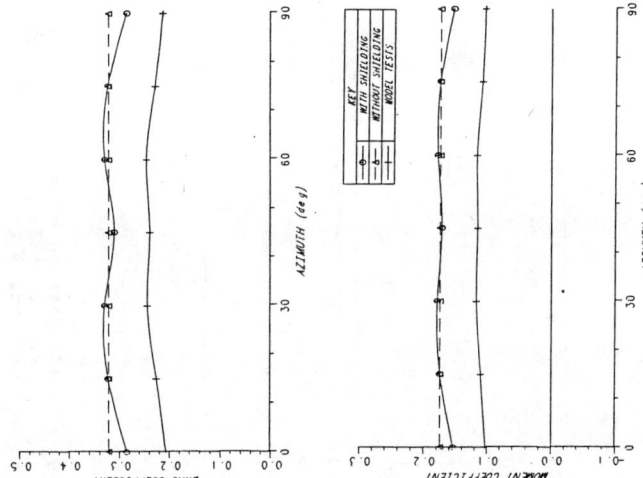

Fig. 11: Comparison of model results with API predictions using methods 3 and 4.

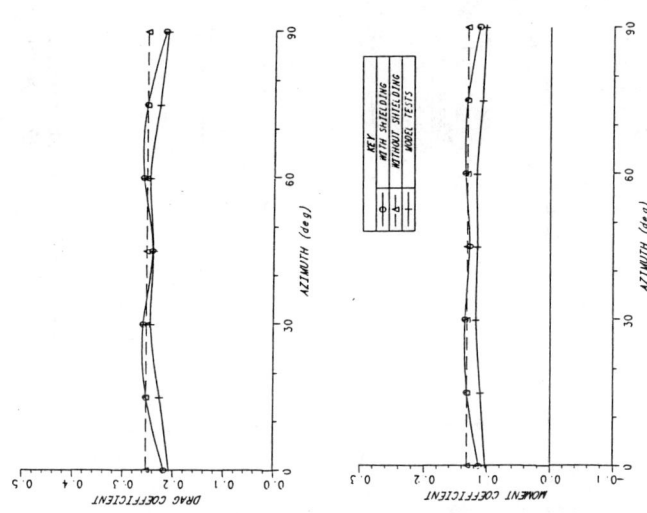

Fig. 10: Comparison of model results with ABS predictions using methods 3 and 4.

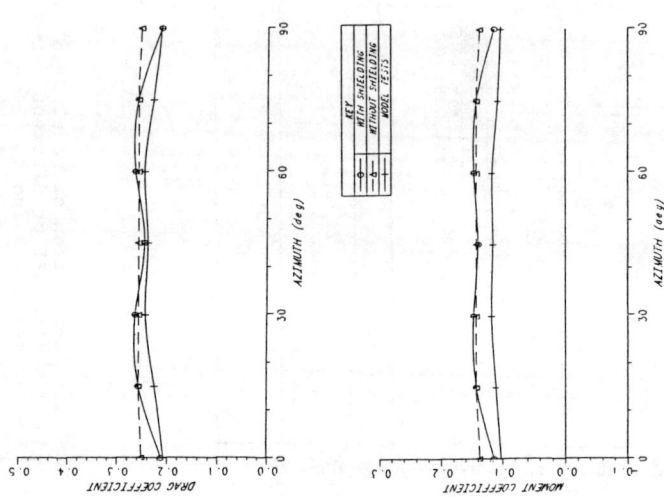

Fig. 13: Comparison of model results with BS 6235 predictions using methods 3 and 4.

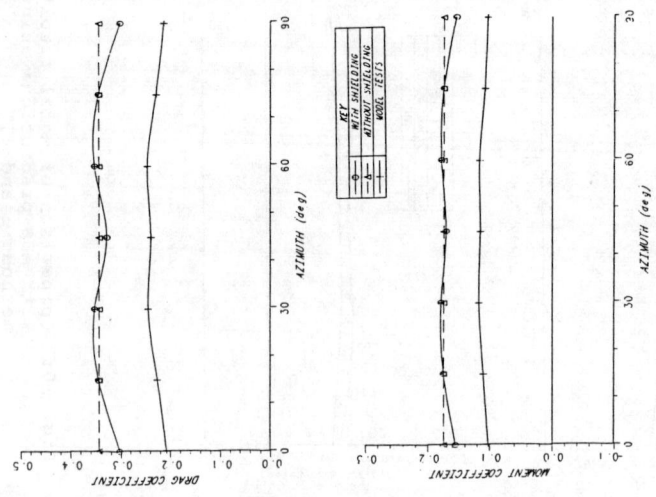

Fig. 12: Comparison of model results with DNV predictions using methods 3 and 4.

LOADS ON DRILLING PLATFORMS 37

TABLE 2: Variation in force and moment coefficient with the position of the model on a static wave profile

Model Position	% Increase in Force	% Increase in Moment
Mid upslope	+12	+11
Crest	-12	- 1
Mid downslope	-10	-10
Trough	+ 5	- 6

More recently some comparative measurements were made of the loads on the deck and superstructure of a semi-submersible tested in a wind-wave facility. The tests were conducted at a linear scale of 1:200 which, for equality of Froude Number, led to a speed scale of 1:14.14. The tunnel has a fetch of 50m (10 km full scale) and at a gradient wind speed equivalent to 43 m/s the naturally generated waves have a peak height of about 6m at full scale. For the purposes of investigating the sensitivity of the wind loads to sea state the naturally generated waves were supplemented by a sinusoidal swell with a height varying from zero to 20m compared to the freeboard of 17m. The drag force coefficients are shown in Fig. 14 for tests in the dry, tests with naturally generated waves and for tests with a superimposed swell of 10m with a period of 17 sec. The dependency of the wind load on swell amplitude is shown in Fig. 15. The differences between the wet and dry tests in the absence of swell are barely significant but there are substantial changes with the addition of a significant low frequency swell. These very limited test results do not permit any conclusions to be reached but it is clear that the influence of the sea-air interface on the magnitude of the wind loads is a problem deserving further attention.

3.0 DYNAMIC WIND LOADS

The low aspect ratio and what might be termed the "untidy" shape of offshore platforms suggests that the integrated dynamic loads will be due primarily to atmospheric turbulence rather than the unsteady wake which is the dominant source of dynamic loading on tall slender buildings and similar structures. This premise was explored by a series of tests conducted on the 1:400 model shown in Fig. 2. The model was tested in scaled turbulent shear flow in which the turbulence level was modified without any appreciable alteration to the spectral shape or the mean speed profile. The turbulence intensities (averaged over the body of the platform) were 12.1%, 15.7% and 21.6%. The influence of intensity on the drag and the lateral rms forces is shown in Figs. 16 and 17. If, as supposed, the load fluctuations are primarily turbulence induced then the rms forces should be roughly proportional to intensity but should tend to a constant if due primarily to wake effects. The former is certainly true although in the case of the lateral forces there is a trend towards a more rapid than linear increase with intensity. The along-wind dynamic forces are dominant and since these combine with the dominant mean load it is the dynamic drag force together with the mean which will produce the critical design case.

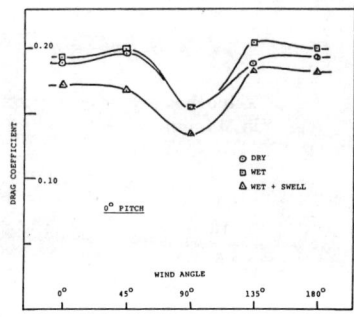

Fig. 14: Variation of deck drag coefficient with the sea-air interface.

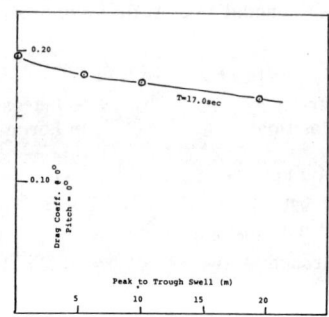

Fig. 15: Variation of deck drag coefficient with superimposed swell.

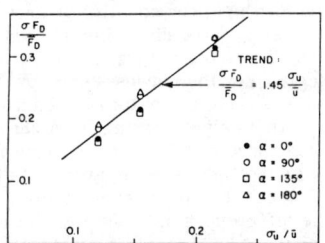

Fig. 16: Variation of RMS drag force coefficient with turbulence intensity.

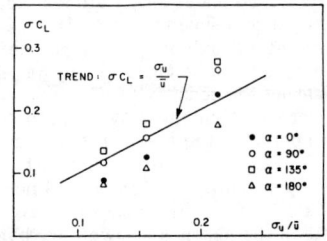

Fig. 17: Variation of RMS transverse force coefficient with turbulence intensity.

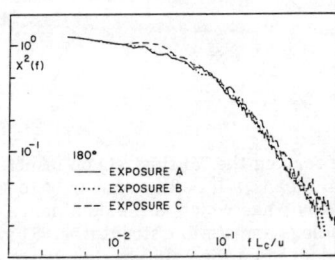

Fig. 18: Drag force admittance functions at $\alpha = 180°$ for exposures A, B, and C (i=.121, .157 and .216).

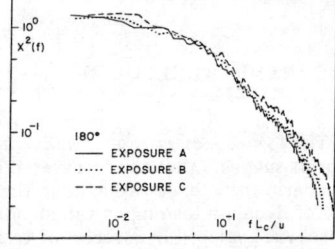

Fig. 19: Drag moment admittance functions at $\alpha = 180°$ for exposures A, B, and C (i=.121, .157 and .216).

LOADS ON DRILLING PLATFORMS 39

The indication of Fig. 16 that the dynamic drag forces are primarily due to turbulence suggests that an aerodynamic admittance function can be defined. The admittance function, $\chi^2(f)$, is defined by;

$$S_F(f) = 4 \left(\frac{\overline{F}}{\overline{u}}\right)^2 \chi^2(f) S_u(f)$$

where,

$\overline{F}$ = mean drag force

$\overline{u}$ = mean speed at the centre of the mean drag load

$S_u(f)$ = longitudinal turbulence spectrum at the centre of the mean drag load

$S_F(f)$ = spectrum of the dynamic drag force.

If the concept of a linear aerodynamic admittance function is indeed correct then it should be invariant with intensity. Figs. 18 and 19 show the drag force and drag moment admittance functions at a wind angle of 180° as a function of $fL/\overline{u}$ where L is a representative dimension of the platform and roughly equal to the square-root of the frontal solid area. For typical values of f, L and $\overline{u}$ of 0.02 Hz, 50m and 40 m/s respectively, the admittance function at the natural frequency of a compliant structure is about 0.9 and is seen to be more or less independent of intensity. It can be concluded that, for compliant structures, the drag force spectrum near the natural frequency can be estimated with acceptable acuracy from quasi-steady aerodynamic assumptions coupled with the assumption of full correlation over the structure.

4.0 CONCLUSIONS

4.1 Mean Loads on the Lena Tower

The extensive tests on the detailed 1:120 and the simplified 1:400 model of the Lena tower demonstrate that despite the complex exterior detail typical of offshore platforms it is possible to design simplified small scale models which yield adequate test data.

4.2 Parametric Studies

The parametric study of models of platforms typical of those of semi-submersibles provided a basis for the assessment of the adequacy of four of the more common rules for determining wind loads. The prime conclusions arising from this study were,

(1) The A.B.S. rules are both simple and adequate for the predictions of mean drag loads at small or zero heel angles.

(2) The API and DNV rules yield significant overestimates of loads.

(3) None of the four are adequate at large angles (> $10°$) of roll or pitch.

(4) Vertical or near vertical forces acting on horizontal or near horizontal surfaces can contribute significantly to the overturning moment.

(5) Shielding effects are more extensive than is suggested by existing design rules and are very significant for members with large diameter legs.

(6) The most satisfactory method of dealing with wind loads when the wind direction is not normal to a face is to compute the load on that face by the application of the usual rules but to use a velocity pressure equal to the component of that pressure in a direction normal to the face in question.

4.3 The Sea-Air Interface

The very limited test program conducted to study the influence of modelling the sea-air interface demonstrated that;

(1) Modelling of the interface by a solid surface appears adequate for small wave heights.

(2) The presence of large low frequency waves have a significant (10% to 20%) influence on the mean loads but further work is required to define the dependency of these influences on wavelength and wave height.

(3) The dynamic wind loads induced by the passage of waves were not studied but tests using static waveforms suggest that these are deserving of attention.

4.4 Dynamic Wind Loads

The studies of dynamic wind loading confirmed that the prime and dominant source of such loading is atmospheric turbulence and that at the frequencies typical of compliant offshore structures it is only slightly conservative to compute the critical along-wind or drag load spectrum near the natural frequency using the assumptions of,

(i) quasi-steady aerodynamics, and

(ii) full correlation over the structure.

ACKNOWLEDGEMENTS

The studies described herein were funded by the Exxon Production Research Company, British Petroleum International and the Natural Sciences and Engineering Research Council (Canada).

REFERENCES

1. Pike, P.J. and Vickery, B.J. "A Wind Tunnel Investigation of Loads and Pressures on a Typical Guyed Offshore Platform", OTC 4288, Proc. 14th Offshore Technology Conference, May, 1982.

2. Vickery, B.J. and Pike, P.J. "An Investigation of Dynamic Wind Loads on Offshore Platforms", OTC 4955, Proc. 17th Offshore Technology Conference, May, 1985.

3. Freathy, P. and Vickery B.J. "Wind Loads on Semi-Submersible Offshore Platforms", OTC 5174, Proc. 18th Offshore Technology Conference, May, 1986.

4. Freathy, P. "Wind Loads on Floating Offshore Structures", M.Eng.Sc. Thesis, The University of Western Ontario, London, Ontario, Canada, August, 1985.

5. Macha, J.M. "An Analysis of Wind Tunnel Data and Calculations Based on Classification Society Methods", Technical Report, Texas Engineering Experiment Station, 1983.

6. American Bureau of Shipping. "Rules for Building and Classing Mobile Offshore Drilling Units", 1983.

7. American Petroleum Institute. "Recommended Practice For Planning, Designing and Constructing Fixed Offshore Platforms", 1984.

8. British Standards Institution. "BS6235 - Code of Practice For Fixed Offshore Structures", 1982.

9. British Standards Institution. "CP3: Chapter 5, Part 2, Code of Basic Data For the Design of Buildings", 1972.

10. Det Norske Veritas. "Rules For the Design Construction and Inspection of Mobile Offshore Structures", 1984.

LABORATORY SIMULATION OF WIND LOADS ON A TENSION LEG PLATFORM

A. Kareem and P. C. Lu
University of Houston,
Houston, TX
and

T. Finnigan and S. L. Liu,
Chevron Corporation, San Ramon, Ca.

ABSTRACT

The mean and fluctuating aerodynamic force and moment coefficients of a typical tension leg platform for various approach wind directions are measured on a scale model exposed to simulated flow conditions in a boundary layer wind tunnel. A parametric study was conducted to determine shielding and interference effects, i.e., the manner in which aerodynamic coefficients are influenced by the location and orientation of the ancillary structures on the platform, e.g., living quarters, flare boom, derricks, etc. The steady aerodynamic coefficients obtained from the classification society recommended procedures provided conservative estimates in comparison with the measured values for all configurations. The results also illustrate that the interference effects among various ancilliary structures on the platform are significant. The specification society procedures do not address the fluctuating unsteady aerodynamic loads, therefore, experimentally derived results provide the only practical means of ascertaining the dynamic load effects.

INTRODUCTION

Wind loads contribute significantly to the overall design loads for tension leg platforms (TLPs). Wind effects on these platforms consist of a mean and a fluctuating component. The mean wind effects result from mean wind force that can be computed from the mean wind velocity and an appropriate aerodynamic force coefficient. The fluctuating component results from the buffeting action of wind gusts or flow-induced effects. The compliant behavior of TLPs in the horizontal plane increases their sensitivity to dynamic effects of wind, which contain a significant level of energy in the low frequency range, that makes them susceptible to dynamic effects of wind. This leads to a relative increase in the overall sensitivity and strength requirements of compliant structures to wind loads compared to those of conventional fixed structures. (1)

The susperstructure of a TLP has a complex geometry consisting of production and drilling equipment, living quarters and a helideck. The information availabe in design codes and specifications to determine aerodynamic force coefficients on offshore platforms has been limited to experimental data obtained from simplified wind tunnel models of basic structural shapes and configurations. Generally the synthesized force coefficient of a platform derived from code recommended values using a projected area approach are conservative. Physical modeling of these platforms, using scale models exposed to simulated atmospheric flow over the oceans in a boundary layer wind tunnel, therefore, continues to serve as a most accurate and practical means of predicting aerodynamic loads.

A program of wind tunnel investigation to determine the mean and fluctuating aerodynamic force coefficients and to better understand the aerodynamics of tension leg platforms was initiated. This paper briefly describes this project.

EXPERIMENTAL PROGRAM

In this study the mean and fluctuating force and moment coefficients of the above-water superstructure of a typical tension leg platform were measured for various approach wind directions. A parametric study was conducted to determine the manner in which these coefficients are influenced by the location and orientation of the ancillary structures on the platform, e.g., flare boom, muhhouse, and derricks. The measured values were compared with force coefficients obtained by synthesising component shape coefficients based on their respective project areas. These comparisions may provide helpful guidance in the identification and subsequent quantification of interference and shielding effects. A flow visualization study was carried out to facilitate a better understanding of the flow field around the platform.

Wind-Tunnel Modelling

The simulation of atmospheric boundary layer winds constitutes an essential prerequsite to any model study of wind-structure interaction. The similitude requirements may be obtained from dimensional arguments derived from the governing equations for fluid motion. A detailed discussion of similarity requirements and their application to wind tunnel testing is provided by Cermak (2). In the study of wind-structure interactions, strong winds are of primary concern. For these winds, thermal stratification of the atmosphere near the surface is destroyed by intense mixing, and the problem becomes one of simulating neutral atmospheric flows which correspond to an isothermal flow in the wind tunnel. The basic requirements for simulating natural winds of this type are:

(1) Undistorted scaling of geometry (geometric similitude),

(2) Reynolds number equality (dynamic similitude),

(3) Rossby number equality (dynamic similitude), and

(4) Similarity of mean velocity and turbulence characteristics (kinematic similitude).

A detailed discussion of these requirements for simlating fluid-structure interaction are given in reference 3.

Wind Tunnel

All measurements were made in the Structural Aerodynamics Laboratory at the University of Houston. The tunnel is an open/closed loop type and is approximately 20 m long and 2m x 1.75m in cross-section. The test section has a movable roof so that satisfactory pressure gradients for all flow conditions can be maintained. An air bearing turntable and remote operation instrument carriage is used for carrying out tests. The turbulent boundary layer is simulated by the natural action of the surface shear over a long fetch of surface roughness on the tunnel floor. The mean speed range from 0.5m/s to 12m/s for the full cross-section with higher speeds for the reduced cross-section.

Wind Velocity Field

The wind velocity at a given point in space is taken to be the sum of the mean wind speed over a suitable period of time and the fluctuating speed described by the standard deviation. As with the flow of any fluid over a surface, a boundary layer is developed in which the wind speed decreases from a maximum value to zero at the surface. A limited amount of data is available from wind measurements taken over the ocean and published results are uncertain, especially when extrapolated to extreme design conditions.

Accordingly, the mean description of the turbulent wind field is therefore assumed to be one of two forms (4,5). The first is the logrithmic law

$$\bar{U}(z) = \frac{\bar{U}_{10} \ln \frac{z}{z_o}}{\ln \frac{10}{z_o}} \tag{1}$$

in which $\bar{U}_{10}$ = reference wind velocity at 10m; z = the vertical coordinate from the mean sea surface; and z_o = roughness length. The second form is an empirical expression represented by a power law

$$\bar{U}(z) = \bar{U}_{10} \left(\frac{z}{10}\right)^a \tag{2}$$

in which a = power-law exponent. The power-law profile has been widely used in many design codes and classification societies because of its simplicity (6). The value of a is dependent on the sea surface roughness and can be expressed in terms of u_* and z_o (4).

The Det norske Veritas (DnV) recommends a = 0.15 for the mean hourly wind velocity (6), whereas American Bureau of Shipping (ABS) recommends a uniform wind speed for every fifty feet segment with a step increase with height from the mean sea surface (7).

The ocean boundary layer was simulated in the structural aerodynamics boundary layer wind tunnel. A set of spires and a horizontal barrier at the entrance of the test section were used in addition to surface roughness to simulate boundary-layer growth. The data for the wind profile was measured using a pitot tube connected to an electronic pressure transducer and a hot film anemometer. The spanwise variation of the wind field at various levels were measured. The variation in the mean wind field over the middle two-thirds of the test section width was less than 1%.

TLP Model

For this investigation two TLP models were utilized. The scales of the models used for measuring steady and unsteady load effects were 1:128 and 1:400, respectively. Only the portion of the TLP above the mean water level was built. The steady load model was constructed out of plastics, brass, and aluminium to avoid unnecessary movements of components during the tests which could lead to cracks. Major componets on the upper deck, such as those identified below, were designed for easy removal to enable testing to be conducted in their absence. However, these components were rigidly attached to the model during the test

   i. drillng derricks and substructures that can be repositioned to represent their range of service to 20 well slots each

  ii. living quarters

 iii. cranes and crane pedestals

   v. flare boom

  vi. utility decks

 vii. pipe, rack, and major components and equipment underneath it.

A simple representation of the lower and mezzanine decks was used by employing a mesh with about 50% porosity, rather than the detailed arrangement of equipment present. The moon pool and the fire walls around the moon pool in the lower and mezzanine decks were included in the model. Figure (1) shows the model with all the ancillary structures on it. This is referred to as base case in this report.

Fig. 1 TLP Model

The model ultilized for the measurement of unsteady load effects was built out of balsa wood and foam. The ancilliary structures were scaled down based on their general configurations. The deck components were removable that permitted a parametric study of the variation in the load effects that result from a change in the deck layout.

Test Configurations

A number of model configurations were tested by modifying or removing components of the basic platform model. These are the test configurations used for steady load measurements.

1. Base case study (Fig. 1), one drilling derricks in northwest location and other in southeast.

2. Base case study same as (1) except move northwest derrick to northeast.

3. Base case as in (2) with northeast derrick removed

4. Base case with both derrick removed.

5. Same as in (4) with cranes and pipe storage removed.

6. Base case with everything removed except, flare bloom, helideck and living quarter.

7. Same as in (6) with living quarters and helideck removed.

8. Base case with everything removed form the platform.

In addition to the above configurations, test runs were made by masking one side of the derrick to represent a wall of pipes that are generally placed during drilling operation. Test runs were also made at different wind speeds and by installing discrete surface roughness to simulate an artificially high Reynolds number flow field around the cylinder. (3)

For the unsteady aerodynamic loads, measurements were conducted for the base case and by removing components similar to the steady measurements but not with the same details. In addition to the unsteady force and moment measurements we also introduced two hot-film sensors, one in front of the model to monitor the far-field turbulence and other in the wake of the model.

Force Transducer and Instrumentation

A special six-component force balance was designed and built for the steady loads study. Due to the large size of the model the force balance was required to be sturdy as well as sensitive to small changes in loads. These characteristics were taken into consideration. Strain gages were used in a number of configurations to

measure six force and moment components. The strain gages were
installed with precision to minimize cross-talk and other spurious
interactions. For the unsteady loads the measurements were made by a
specially developed ultra-sensitive five-component high-frequency
force transducer. The outputs of the transducers were sent through a
signal conditioning unit and amplifiers before being fed into the
analog-to-digital converter. The analog-to-digital converter was
interfaced with a PDP Micro-11 computer for data acquisition and
reduction (8).

Results and Discussion

The aerodynamic force and moment coefficients for the tested
configurations are plotted as functions of the approach wind
direction in Figs. 2a and 2b. In Fig. 2a the drag coefficients have
maximum values for both base cases (configuration 1 and 2) and
minimum values for the platform with all the ancillary structures
removed (configuration 8). The plotted values exhibit a trend of
successive valleys and peaks with maximum values for the wind
directions along the diagonals and minimum values for wind direction
perpendicular to one of the sides. This trend can be attributed to
the fact that the drag coefficient for all wind directions were
obtained by normalizing with the same reference area. However, the
frontal area changes with the wind direction, leading to maximas for
wind approaching along diagonals that contributes maximum projected
area with maximum associated aerodynamic force. The difference in
the values of the drag coefficient for the two base case studies
indicates that interference and proximity effects do influence the
aerodynamic force. For wind approaching at zero degree angle of
attack, the base case corresponding to configuration 1 had a higher
drag coefficient in relation to configuration 2. Although both
configurations are comprised of the same ancillary deck structures
with the same configuration except for the location of two drilling
derricks (Fig. 1), in configuration 1 both derricks were exposed to
wind for zero angle of attack, whereas, in configuration 2 the south
side derrick was located in the shadow of the front derrick which
resulted in an overall reduction in the drag as exhibited in Fig. 2.
There is a concomitant decrease in the drag coefficient as the deck
structures are removed corresponding to configurations 5, 6, 7 and 8
leading to a basic platform superstructure without ancillary
components.

The drag-induced overturning moment or pitching moment is
plotted as a function of wind direction in Fig. 2b and generally
follows the variation of the drag force with the wind approach angle.
The measured pitching moment includes contributions from both direct
aerodynamic sources, e.g., moment induced by the drag force, as well
as the flow-induced effects, such as lift forces resulting from flow
separation. The flow separation takes place at the leading edge of
the deck and is more predominant at the living quarters and the
helideck. In the absence of a leading edge for separation the lift
induced is less effective due to several deck components obstructing
the flow. The unsymmetrical distribution of lift induced at the
leading edge surface tends to enhance the overtuning moment. It is

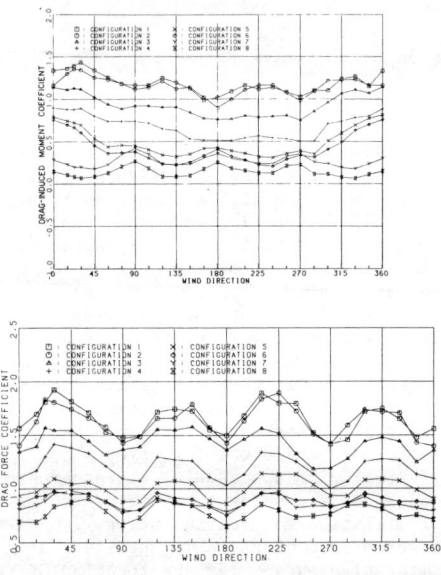

Fig. 2a-Drag Force Coefficients
   b-Drag-induced Moment Coefficients

flow-induced load effects such as the lift-induced pitching moment which cannot be estimated using code specifications and often are neglected in the design procedure. These effects may be seen by examining the values of the drag-induced moment at wind approach angles of 0 and 180 degrees. More detailed discussion of the results reported here and the force and moment coefficients in other directions are given in reference 3. Also included in reference 3 are the results of measurements made to investigate the dependence of measured force and moment coefficients on Reynolds number and the influence of employing discrete roughness elements at selected locations, on the circumference of each leg, on the Reynolds number invariance.

COMPARISON OF METHODS

The experimentally-derived force and moment coefficients were used to validate the estimated values of these coefficients based on the projected area approach. The synthesis of the overall force and moment coefficients was carried out in accordance with the procedures recommended by the American Bureau of Shipping (ABS) and Det Norske Veritas (DnV), although other societies have quite similar guidelines.

The drag force and drag-induced moment (pitch) were computed, for all the TLP configurations tested in the wind tunnel experiment for wind approaching from the north, east, northeast, and northwest, using the ABS and DnV specifications.

The typical comparison with the experimental values is shown in Fig. 3. In a majority of the configurations, the DnV estimates are notably higher than the ABS estimates and the wind tunnel results. This is particularly true for the base case configurations. The ABS estimates are in almost every configuration lower than the corresponding DnV estimates.

The estimated coefficients of drag-induced moment were lower than the corresponding wind tunnel measurements for configurations in which the drilling derricks had been removed but the helideck and the living quarters were still in place. In these configurations the flow-induced aerodynamic lift contributes to the pitching moment. As discussed earlier, the projected area approach fails to include this contribution which leads to the underprediction of the drag-induced moment.

Unsteady Loads

The data obtained from unsteady measurements is being analysed. This includes simultaneously monitored five components of aerodynamic loads as well as velocity fluctuations in the far-field and in the wake. This data will lend itself to the development of an aerodynamic admittance function. In addition to these measurements an extensive mapping of the space-time fluctuations in the wind field have been made. These data will facilitate the development of a theoretical model, based on a quasi-steady and strip theories, for

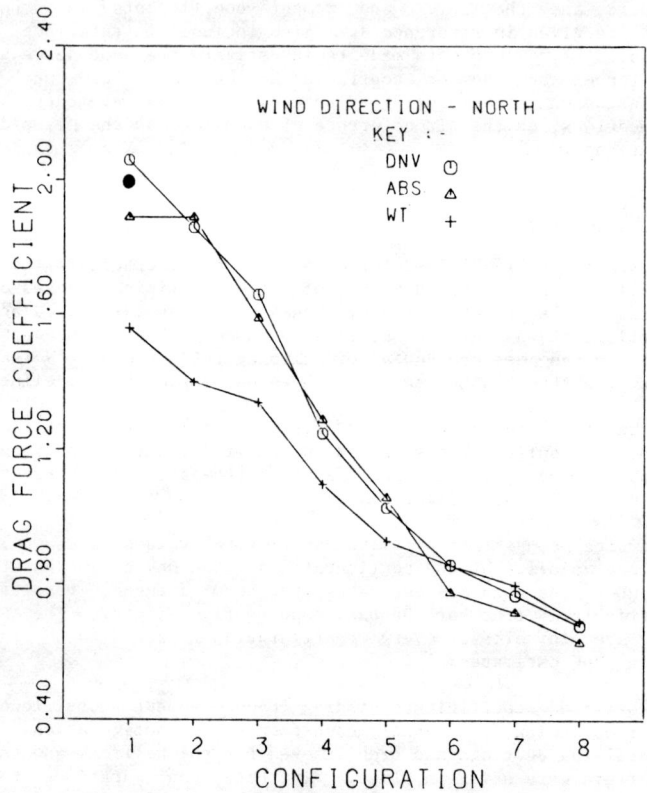

Fig. 3 Comparison of Predicted and Measured Drag Coefficients

ascertaining dynamic load effects on TLPs.

The preliminary spectral analysis of the pitching moment suggests the presence of additional energy at high frequencies. This may introduce higher fatigue loads on the tendons of a TLP since the damping available from both structural and hydrodynamic sources is very low in this mode of vibration. The coherence estimates of the space-time velocity fluctuations exhibit departure from commonly accepted form of the coherence functions used in practice for wind sensitive structures. A discussion of the influence of the sea surface on the space-time correlation of the wind velocity field is available in references 4 and 9. Some of the data pertaining to the unsteady load measurements, that is presently being analyzed, will be discussed during the presentation of this paper.

CONCLUSIONS

A wind tunnel investigation of the aerodynamic loads on a typical tension leg platform has led to the following major conclusions

- a. The current classification society procedures based on projected area approach tend to overestimate force and moment coefficients.

- b. Both aerodynamic load measurements and flow visualization experiments suggest that flow-induced lift forces over the helideck and living quarters contribute significantly to the drag-induced moment on the platform at the mean sea surface.

- c. Interference and proximity effects among various ancillary deck structures are notable and influence the aerodynamic force and moment coefficients.

- d. The drilling derricks introduce prominent wind load effects in terms of their contribution to the overall aerodynamic loads.

- e. The preliminary analysis of the unsteady loads suggests the presence of additional energy in the pitching moment spectrum at high frequencies.

ACKNOWLEGEMENTS

The support for this research was provided by the Chevron Corporation and the NSF-PYI-84 award to the senior author by the National Science Foundation under Grant No. CEE8352223. Any opinions, findings, conclusions or recommendations expressed in this publication are those of the authors and do not necessarily reflect the views of the sponsors.

REFERENCES

1. Kareem, A., "Wind-Induced Response Analysis of Tension Leg Platforms," J. of the Struct. Div., ASCE, Vol. 111, No. 1, January, 1985.

2. Cermak, J.E., "Laboratory Simulation of the Atmospheric Boundary Layer," AIAA, Vol. 9, No. 9, Sept. 1971.

3. Kareem, A. and Lu, P. C., "A Wind Tunnel Investigation of Aerodynamic Loads on a Typical Tension Leg Platform" Dept. of Civil Engineering, University of Houston, Report NO. UHCE 85-9, December 1985.

4. Kareem, A., "Structure of Wind Field Over the Ocean," International Workshop on Offshore Winds and Icing, Halifax, Nova Scotia, Atmosphereic Environment Service of Canada, Oct. 1985.

5. Simiu, E. and Scanlan, R. H., "Wind Effects on Structures," An Introduction to Wind Engineering, Wiley Interscience.

6. Det Norske Veritas, Rules for the Design Construction and Inspection of Offshore Structures, Appendix B, 1982.

7. American Bureau of Shipping. Rules for Building and Classing Mobile Offshore Drilling Units, 1980.

8. Kareem, A., Lu, P.C., and He, T., "Software for Automated Experimental Control, Data Acquisition and Reduction on a PDP-Micro 11," University of Houston, Department of Civil Engineering Technical Report, No. UHCE 85-10, Jan. 1985.

9. Vickery, B. J. and Pike, P.J. "An Investigation of Dynamic Loads on Offshore Platforms," 17th Offshore Technology Conference, OTC 4955, 1985

A WIND/WAVE TUNNEL STUDY OF COMBINED WIND AND WAVE LOADS ON A TLP

P. J. Vickery[1] and A. G. Davenport[2]
Boundary Layer Wind Tunnel Laboratory
The University of Western Ontario
Faculty of Engineering Science
London, Ontario, Canada
N6A 5B9

Introduction

In recent years the offshore oil industry has began developing oil fields located in deep waters. In these locations it is not economical to employ the conventional, relatively rigid fixed type of platform commonly used for shallower water applications. The newer deep water platforms are designed to be compliant and move with the waves as opposed to the shallow water structures which resist the wave forces through their relatively high stiffness. The compliant structures have very long surge and sway fundamental periods of vibration (of the order of 100 seconds) and are therefore not particularly sensitive to the first order wave forces which typically have most of their energy associated with periods of 20 seconds or less. These long periods of vibration do however make complaint structures sensitive to second order wave drifting forces and turbulent wind. These long periods of vibration tend to coincide with the range where the energy of the fluctuating wind is a maximum, and consequently wind forces can contribute significantly to the motions of compliant platforms.

Until recently, wind loads were treated in a rather simplistic manner, usually using a static drag load without any allowance for resonant amplification effects, unbalanced loading torsion etc.; at the same time a significant effort was put into a complex and refined description of the wave loading problem. This approach is generally satisfactory for rigid structures such as jacket type towers but is not satisfactory for structures where torsional (yaw) response may be produced by unbalanced wind loading (eg. some jack-up rigs (1)) or particularly low frequency compliant structures (2) such as tension leg platforms.

In the case of tension leg platforms the wave response is dominated by second order slowly varying drift forces and there is no clear indication as to how the low frequency wind will affect this response. This question, and others, need to be addressed experimentally using models tested in a facility which combines both the wind and wave forces at the same time. Such a program is currently in progress at the Boundary Layer Wind Tunnel Laboratory at the University of Western Ontario where simple models of tension leg platforms are being examined in a wind-wave tank capable of simulating the action of random sea states and turbulent wind.

---

1. Research Assistant, Boundary Layer Wind Tunnel Laboratory, The University of Western Ontario, Faculty of Engineering Science, London, Ontario, Canada, N6A 5B9

2. Director and Professor, Boundary Layer Wind Tunnel Laboratory, The University of Western Ontario, Faculty of Engineering Science, London, Ontario, Canada, N6A 5B9.

Wind Loading – Current Approaches

An experimental approach used recently (2,3,4,5) consists of a rigid model of a platform mounted on a rigid balance system which measures base shears, moments and torques and placed in a boundary layer wind tunnel. This approach allows for an accurate reproduction of the prototype so as to include all details of the derricks, living quarters, helicopter landing pads etc. all of which contributed to the moments, shears and torque in a very complex manner. The mean measured forces and their spectral density functions can then be combined analytically with estimates of the hydrodynamics loads (obtained either theoretically or experimentally) to determine the total response of the structure to the wind and wave loads.

The windfield in which the model is tested should be representative of the full-scale wind approaching over a deforming boundary (the sea surface). There is unfortunately a shortage of full scale measurements of wind flow over water.

Another experimental approach occasionally used to account for wind loads does not use a boundary layer wind tunnel but rather a bank of fans located on one side of a wave tank (see for example, reference 6), while this approach does have the advantage of combining a desired wave field with a wind field, it is unlikely that a realistic simulation of a natural wind will result.

A theoretical (or empirical) approach has been used, for example, by Kareem (2,4) and Simiu (5) among others. This method uses a quasi-steady drag type analysis for the wind loading and typically a Morrison equation wave load model.

For the wind forces the frontal area of the structure is broken down into many small elements such that the mean induced surge force can be described by:

$$\overline{F} = \int_A v_2 \rho C_D \overline{U^2} (y,z) \, dA \tag{1}$$

and the spectral density of the fluctuating component of the surge force is expressed as:

$$S_{FF}(f) = \overline{U}^2_{ref} S_u(f) \chi^2(f) \tag{2}$$

where $\chi^2(f)$ is an aerodynamic admittance function accounting for the degree of correlation associated with the fluctuating wind at different locations on the structure, and $S_u(f)$ is the spectral density of the longitudinal velocity obtained at a reference position on the structure.

Once $\chi^2(f)$ and $S_u(f)$ are defined it is then possible to combine the wind force spectrum with the wave force spectrum to determine the response of the structure. The response may be calculated in either the frequency domain or the time domain. The frequency domain approach is simpler than the time domain approach however there are non-linearities present in both the wave-induced drag forces and the restoring forces of a TLP which must be linearized and/or ignored. The time domain solution is often approached using a Monte Carlo type simulation of the wind and wave forces so that

$$F_{wind}(t) = \sum_{j=1}^{n} F_j \cos(2\pi f_j t + \phi_j)$$

where ϕ_j is obtained from uniform randomly distributed numbers between 0 and 2π. A similar expression is used for the wave forces which is integrated over the instantaneously submerged portion of the structure at each time step. The equation of motion is then solved using a suitable numerical differential equation solver such as Runge-Kutta. This approach enables all non-linearities to be included but is computationally time consuming and expensive.

It should be noted that the quasi-static approach to the wind loading problem ignores interference effects between the many components of the TLP which are known to produce significant differences between calculated and measured values of mean wind induced forces and $\chi^2(f)$ and that wind tunnel testing of a model will produce more reliable results.

Velocity Spectra

An important question in the evaluation of the response of compliant offshore structures is the definition of the energy in the wind spectrum at the structures' very low natural frequencies. Models for the spectrum of energy in the natural wind over land such as those proposed by Davenport and Harris may not fully reflect the variability present at these low frequencies.

Spectra of longitudional velocity in high winds measured on a jack-up rig off the coast of Sable Island near Nova Scotia (7) showed a remarkable range of intensities at these low frequencies. This range was strongly associated with wind direction and this in turn could be linked either to the presence of the low island in the one direction, open sea in the other or to differences in atmospheric stability.

Wind-Wave Tunnel Testing

Previous wind tunnel studies carried out on offshore structures in boundary layer wind tunnels designed for testing land base structures have not incorporated any dynamic wind-wave-structure interaction. This may play an important role in the response of the structure.

There are several possible sources of interaction coupling the dynamic responses to the wind and waves. These include first the influence of the movement of the dominant waves whose velocity is close to that of the wind introducing a moving boundary effect; second the changes in the silhouette area of the structure as the trough and crest move past the structure, and last the contribution of hydrodynamic damping in controlling wind-induced motion.

Testing offshore platforms, in a wind-wave facility allows for all of the above wind-wave-structure interaction terms to be accounted for and provides the most reliable means of validating any analytical model which includes both wind and wave forces. Such a facility is currently available at the University of Western Ontario. The wind-wave tank consists of a 52 metre long tank 2 metres deep and 5 metres wide. The 52 m fetch enables the natural generation of a short crested sea which is supplemented by longer waves produced from a wave generator at the end of the tank. Through the use of an on-line computer-controlled feed-back system, a two dimensional sea state can be generated as prescribed by the user. The maximum attainable wind speed in the wind-wave tank is approximately 10 m/s.

Research on the effect of wind-wave-structure interaction includes testing simple models of tension-leg platforms in the wind-wave tank as well as in the standard 'dry' boundary layer wind tunnel. The testing procedure involves instrumenting tethers on the TLP's in order to measure changes in tension and to monitor the platforms motion in all 6 degrees of freedom (3 translations, 3 rotations).

It is hoped that this research will provide a vehicle for improving the current methods of incorporating wind loads into the design of tension leg platforms and other compliant offshore structures.

References

1. Davenport, A. G. and Vickery, B. J., "Wind Loading of Off-Shore Structures", First Canadian Conference on Computer Methods in Offshore Engineering, Halifax, Nova Scotia, May 1984.

2. Kareem, A., "Non-Linear Dynamic Analysis of Compliant Offshore Platforms Subject to Fluctuating Wind", Journal of Wind Engieneering and Industrial Aerodynamics, 14, 1983 pp. 345-356.

3. Armstrong, B. J., Barnes, F. H., Drabble, M. J. and Grant, I., "Unsteady Aerodynamic Loading at a Tension Leg Platform", Offshore Technology Conference, 4641, May 1983.

4. Kareem, A. "Wind-Induced Response Analysis of Tension Leg Platforms", ASCE , Journal of Structural Engineering, Vol. 111, No. 1, January 1985.

5. Simiu E. and Leigh S. D., "Turbulent Wind and Tension Leg Platform Response", ASCE, Journal of Structural Engineering, Vol. 110, No. 4, April 1984.

6. Faltinsen, O. M., Fylling, I. J., Teigen, P. S., Van Hooff, R. W., "Theoretical and Experimental Investigations of Tension Leg Platform Behaviour", Proceedings, Behaviour of Off-Shore Structures (BOSS 82') Cambridge, Massachussets, 1982.

7. Vickery, P. J., Vickery, B. J. and Davenport, A. G., "Gust Factors and Turbulence Measured at 50 m Over the Sea Off Sable Island", Proceedings, International Workshop on Offshore Winds and Icing, Halifax, Nova Scotia, October 1985.

Amplification of Wind Effects on Compliant Platforms

Graham R. Cook*M. ASCE, Thusitha Kumarasena*,
and Emil Simiu**M. ASCE

A brief review is presented of recent information on the estimation of hydrodynamic damping. Results of simplified calculations are then presented, which indicate that typical tension leg platforms do not appear to experience significant amplification of wind-induced dynamic surge motions. This is the case not only for platforms in extreme environments, for which earlier investigations have been reported, but for non-extreme environments as well. It is noted that smaller values of the drag coefficient in the Morison equation result in increased amplifications of the fluctuating wind-induced response. However, such increases are in most cases amply compensated by the smaller hydrodynamic exciting forces associated with these smaller values. Thus, it appears that the use of a large value of the drag coefficient usually is conservative from a structural design viewpoint. It is also pointed out that data concerning drag coefficients corresponding to the large Reynolds numbers (of the order of one million or more) and small Keulegan-Carpenter numbers (of the order of unity or less) of interest in tension leg platform design do not appear to be available in the literature; in the present state of the art such data cannot be obtained from laboratory tests and should be inferred from full-scale measurements.

Introduction

The natural frequencies of surge motions in compliant offshore structures such as tension leg or guyed tower platforms are quite low (of the order of 0.01 Hz), while frequencies of ocean waves have orders of magnitude of 0.1 to 1 Hz. Hydrodynamic forces having the same frequencies as the ocean waves (i.e., first order wave forces) therefore do not produce dynamic amplification effects in surge. This is an important advantage of compliance. However, second

*Graduate Student, Dept. of Civil Engineering, The Johns Hopkins University, Baltimore, Md.

**Research Engineer, National Engineering Laboratory, National Bureau of Standards, and Research Professor, Dept. of Civil Engineering, The Johns Hopkins University, Baltimore, Md.

order wave forces (due to nonlinear hydrodynamic effects), as well as wind forces, can have significant components with frequencies of the order of 0.01 Hz. The question therefore arises whether these forces can produce significant dynamic amplification effects.

In this note we confine ourselves to discussing dynamic effects due to wind loads. Their severity depends upon the magnitude of the fluctuating wind loads acting on the platform, and on the damping available in the fluid-body system that consists of the platform oscillating in the ocean flow. Wind-induced loads can be estimated fairly simply using empirical drag coefficients available from existing or ad-hoc wind-tunnel studies, in conjunction with expressions for the relation between wind pressures and wind velocities and for the spectral and cross-spectral density of the along-wind turbulent velocity fluctuations in atmospheric flows (see, e.g., [1, Sects. 14.2 and 2.3] and [3]). On the other hand, in the current state of the art, the estimation of damping can pose difficult problems. Since the damping is overwhelmingly of hydrodynamic origin, it can be stated that the estimation of the wind-induced dynamic response of compliant structures of the tension leg or guyed tower type is primarily a hydrodynamics problem.

The purpose of this note is to review briefly recent information on the estimation of hydrodynamic damping, and to present and discuss results of simplified calculations aimed at estimating the susceptibility of a typical tension leg platform to wind-induced dynamic effects. The emphasis in this work is on non-extreme environments, which have not been investigated previously, and in which the amount of hydrodynamic damping available in the fluid-body system is relatively reduced.

Hydrodynamic Damping

The fluid-body interaction produces waves that travel away from the body. The so-called radiation damping is associated with energy expended in this process. For low-frequency motions (0.01 Hz, say), the radiation damping appears to be negligibly small and for this reason it has not been taken into account in calculations of dynamic wind-induced response [2,3].

The only significant source of hydrodynamic damping is, then, the fluid viscosity. This is accounted for through the drag force term in the Morison equation, which is written in the form

$$D = 0.5 \rho \, C_d \, A \, |V - \dot{x}| \, (V - \dot{x}) \qquad (1)$$

where A = projected area of cylinder on a plane normal to

the flow, V = horizontal velocity of the fluid, $\dot{x}$ = horizontal velocity of the body, and Cd = drag coefficient.

The drag coefficient depends on the Keulegan Carpenter number (K) and, as first shown by Sarpkaya, on the Reynolds number (Re) as well. It can be verified that, for an 18m diameter column of a tension leg platform, and assuming the waves to be harmonic, K and Re have the approximate values listed below:

	H = 25m; T = 18s		H = 3m; T = 6s	
	K	Re	K	Re
z = 0	4	70×10^6	0.5	30×10^6
z = 30m	3	50×10^6	0.02	1×10^6

(H = wave height, T = wave period, z = depth below mean water level.)

Until recently very little information was available in the literature on drag coefficients Cd for small values of K (about 4 or less). As noted in [4], "practically all the laboratory and ocean-based experiments have been conducted for K larger than about 4 and it is assumed that Cd for K < 4 is unimportant..." Graphs of the evolution of Cd with K given for circular cylinders in [5] and [6] suggested a monotonic decrease of Cd with decreasing values of K from a local maximum corresponding to a critical value $K^* \simeq 8$ to 12 (depending on the value of Re) to markedly lower values for K $\simeq$ 4 or so.

Experiments reported in 1985 by Bearman et al. [7] showed that Cd decreases monotonically as K decreases from the critical value K^* to K $\simeq$ 2 to 3 or so (depending on Reynolds number). For values K < 2 to 3 it was observed that Cd increases as K decreases. For these values the drag force was attributed in [7] to skin friction and boundary-layer displacement effects. Typical values reported in [7] for K $\simeq$ 2 are: Cd $\simeq$ 1.0 for K $\simeq$ 2, Re $\simeq$ 1000; Cd $\simeq$ 0.6 for K $\simeq$ 2, Re $\simeq$ 3000; Cd $\simeq$ 1.8 for K $\simeq$ 0.6, Re $\simeq$ 1000. Note, for a given value Re, the increase of Cd as K decreases. Note also, for a given value K, the decrease of Cd as Re increases.

In 1986 Sarpkaya reported test results that made possible a more refined assessment of the hydrodynamic loads for low values of K, and identified three flow regimes in the range 0 < K < K^* [4]. For 0 < K < Kc the flow is laminar, attached, and stable. This is referred to in [4] as the Stokes-Wang regime [8]. (For Re = 780, Kc $\simeq$ 0.75; as Re becomes larger, Kc decreases). In the Stokes-Wang regime expressions derived from first principles show that Cd increases as K decreases [4, 8]. Experimental values of Cd reported in [4] were found to agree closely with the values yielded by these expressions.

For $K_c < K < K_u$, owing to instability phenomena in the boundary-layer, the flow is no longer two-dimensional. (For Re = 1700, $K_u \simeq 1.6$) Rather, it exhibits axially periodic mushroom-shaped vortices, first observed by Honji [9]. A stability analysis motivated by Honji's observations was presented by Hall [10]. At and near the onset of the instability the flow remains laminar. However, as K increases further, the flow separates from the cylinder surface, and transition to turbulent flow occurs at the value $K = K_u$. The data of [4] suggest that the minimum value of Cd occurs at this value. From the data of [4] it appears that the minimum value of Cd decreases as the Reynolds number increases. (For Re = 1800, $Cd(K_u) \simeq 0.8$; for Re = 39000, $Cd(K_u) \simeq 0.5$.) Over most of the interval $K_c < K < K_u$, as K decreases Cd increases from its minimum value $Cd(K_u)$.

For $K_u < K < K^*$ the influence of flow separation and vortex shedding becomes increasingly significant, and Cd increases with increasing K.

From an engineering point of view, it is of interest that Cd can have substantial values (about 1.0 or even 2.0) for small K. However, the results of [4] and [7] were obtained in the laboratory and could therefore not provide information on drag coefficients corresponding to the very large Reynolds numbers of interest in tension leg platform design. Such information can only be obtained from full-scale tests. It may well be that Cd values corresponding to small K values and to Re values of the order of one million or more are quite small, perhaps of the order of 0.1 or even less. If it is conjectured that the drag coefficient corresponding to $K = K_u$ is a minimum not only for Re values such as those reported in [4] and [7] but for very large values of Re as well, then an upper bound for that minimum value can be estimated by using the following expressions:

$$K_c = 5.8 \, (Re/K_c)^{-1/4} \, [1 + 0.205 \, (Re/K_c)^{-1/4}] \quad (2)$$

[10, pp.356 and 348, and 4, p.63], and

$$Cd(K_u) < Cd(K_c) \simeq 46.5 \, K_c^{-1} \, [0.56(Re/K_c)^{-1/2} + 0.32/(Re/K_c)] \quad (3)$$

[4, p.62]. For example, for $Re = 10^7$, $K_c \simeq 0.05$, and $Cd(K_u) < 0.05$.

Such estimates are tentative, since neither the previously stated conjecture on the relative magnitudes of

Cd(Ku) and Cd(Kc), nor the applicability in this context of Eqs. 2 and 3 is clearly established. In addition, the assumption that the cylinder is smooth, on which most of the work just reviewed is based, may not hold in practice owing to the possible presence of marine growth on the surface of platform members.

In the absence of dependable information on the hydrodynamic damping available at very large Reynolds numbers, it is useful to perform exploratory studies aimed at assessing the effect of uncertainties with respect to hydrodynamic damping on the wind-induced dynamic response of compliant platforms. Laboratory tests of structures such as tension leg platforms violate Reynolds number similarity requirements by several orders of magnitude and, as far as the hydrodynamic damping in the full-scale situation and its effects on the response are concerned, can yield grossly misleading results. It is therefore necessary to perform such studies numerically. This is done subsequently in this note for a number of representative loading situations in the case of a tension leg platform (TLP).

Estimates of Wind-Induced Dynamic Response in Surge for a Tension Leg Platform

The effects of wind fluctuations on tension leg platforms (TLP's) have been studied in [3] for extreme wind, wave, and current conditions using a time domain approach. It was concluded in [3] that, under such conditions, the dynamic amplification of the fluctuating action of wind is relatively small. This is due to the significant energy dissipation that occurs during the surge motion induced by the extreme loading conditions assumed in [3]. Such dissipation is accounted for through the drag term in the Morison equation (Eq. 1), and increases in a complex, nonlinear fashion as the relative fluid-structure motions increase. It was shown in [3] that dynamic amplifications of wind-induced surge motions under extreme conditions were relatively small even if the drag coefficient was assumed to be as low as Cd 0.1 (i.e., the amplifications were in effect equivalent to those occurring in a linear oscillator with a damping ratio $\zeta \simeq 0.25$ or more, see [3, p. 796]). Note that in studies of wind-induced response reported in [2] it was assumed that Cd has considerably higher values, e.g., Cd = 0.8.

The question arises whether dynamic amplifications of wind effects remain relatively small when the platform moves under the action of less than extreme environments. The purpose of this note is to present an investigation into this question using the time-domain approach described in [3].

The models used in this note for the structure of the wind and ocean environment, the aerodynamic and hydrodynamic loads, and the platform behavior, are the same as those of [3], to which the reader is referred for details. Note that the hydrodynamic loads in [3] and in this note are based on simple models proposed in [2]. More elaborate modeling is possible, in particular with respect to the effect of current on the drag force, as suggested in [2] and [11]. However, this was not judged necessary for the exploratory purposes of this note. The TLP mass and geometry (with 18m diameter columns and 9m diameter pontoons) and the tendon numbers, dimensions, and prestress were also assumed to be the same as in [3]. The water depth was assumed to be 550m. The following sets of Morison equation coefficients were used in the calculations: $Cd = 0.62$, $Cm = 1.8$, and $Cm = 0.1$, $Cm = 1.8$. In all cases being considered calculations were carried out by assuming, first, the presence of current with speed varying linearly from 1.4m/s at the mean water level to 0.15m/s at the bottom and, second, the absence of current. Turbulent longitudinal wind speed fluctuations corresponding to mean hourly wind speeds at 35m above mean water level $U(35) = 45m/s$, 30m/s, and 15m/s were generated by Monte Carlo simulation from spectral expressions given in [3]. A time history of the longitudinal wind velocity fluctuations for $U(35) = 30m/s$ is shown in Fig. 1. For the purposes of this note, waves were assumed to be monochromatic with heights $H = 25m$, 12m, and 6m for the 45m/s wind; $H = 18m$, 12m, 6m, and 3m for the 30m/s wind; and $H = 12m$, 6m, and 3m for the 15m/s wind.

To assess the extent to which the hydrodynamic damping is effective in preventing the occurrence of large dynamic amplifications of the wind-induced surge, the platform response was also calculated for two idealized wind loading cases. In the first case the wind speed consisted of the superposition of the mean speed $U(35)$ and of harmonic fluctuations with period equal to the nominal natural period of the platform, the amplitude of the fluctuations being equal to the rms of the turbulent fluctuations which would correspond in natural winds to that mean speed. The wind load caused by this assumed wind speed is denoted by $W(t)$. This case allowed estimates to be made of an equivalent linear damping ratio, i.e., of the damping ratio of a linear dynamic system having the same total mass (including the added mass) and spring constant as the platform, and experiencing under the action of the wind load, $W(t)$, a peak surge motion Sw, where Sw is equal to the peak of the nonlinear wind-induced surge motion of the platform subjected to both the wind load $W(t)$ and the nonlinear hydrodynamic loads (see [1,3] for details).

The second idealized wind loading case consisted of a wind speed assumed to be constant in time and approximately equal to the highest one-minute wind speed corresponding to the specified mean hourly speed, i.e., the wind loading was assumed to be induced by a constant wind speed equal to 1.25 x $U(35)$. A comparison between the response of the platform subjected to the hydrodynamic loads and this idealized load on the one hand, and to the hydrodynamic loads and the actual (simulated) turbulent wind load on the other, allows a conclusion to be drawn on the severity of the wind-induced fluctuations in terms of a conventional, standard loading situation -- the one-minute wind load.

Consider, for example, the surge response record shown in Fig. 2, for which U(35) = 15m/s, H = 3m, current was assumed to be present, Cd = 0.62, Cm = 1.8, and the wind speed fluctuations are harmonic with period T = Tp = 100s, where Tp = natural period of platform. The equivalent linear damping ratio, based on the amplitude of the wind-induced fluctuations (i.e., of the fluctuations with period T = 100s in Fig. 2), was in this case determined to be ζ = 0.3. Figure 3 shows the surge response corresponding to the same conditions as for Fig. 2, except that the wind speed fluctuations are now turbulent. It is seen that the amplification of the wind speed fluctuations is in this case of the same order or smaller than that corresponding to the harmonically fluctuating wind. Figure 4 shows the surge response corresponding to the same conditions as Figs. 2 and 3, except that the wind speed is now constant and equal to 1.25 x 15 = 18.75m/s. In this case the peak steady state response is seen to be somewhat smaller in turbulent wind than under the effect of the constant one-minute wind. Figure 5 shows the response in turbulent wind with U(35) = 15m/s, Cd = 0.62, Cm = 1.8, assuming no current and no waves (clearly a purely hypothetical case). It is seen that the dynamic amplification of the dynamic effects of wind is considerably larger in this case than in Fig. 3; nevertheless, this is more than compensated by the fact that the exciting hydrodynamic force for Fig. 5 is reduced (in this particular case to zero), that is, the total response is smaller in Fig.5 than in Fig. 3. Figure 6 shows the response of the platform corresponding to the same conditions as in Fig. 3, except that Cd = 0.1 instead of Cd = 0.62. A reduction in the value of Cd from 0.62 to 0.1 has again two opposite effects: an increase in the amplification of the fluctuating motions induced by wind, and a decrease in the hydrodynamic exciting force. It was found that, in the cases examined in this work, the latter effect was considerably stronger than the former so that, in spite of the increased dynamic amplification of wind effects, the use of a large value of Cd usually is conservative from a structural design viewpoint. This is illustrated by a comparison between Figs. 3 and 6.

Conclusions

The calculations carried out in this work indicate that the dynamic amplification of wind-induced effects increases with decreasing current and waves and with decreasing values of the drag coefficient Cd in the Morison equation; however, even for values Cd as small as 0.1 the calculated peak response under turbulent winds did in most cases not exceed the response calculated by assuming the wind speed to be constant and equal to the highest one-minute wind. In the few cases in which it did (which corresponded to wave heights H = 3m), the error inherent in this assumption was relatively small (less than about 15 percent).

It thus appears that the dynamic amplification of wind-induced surge motions is not a major factor for environmental conditions of practical interest. However, the validity of this conclusion is not necessarily general. The assessment of the possible susceptibility to such amplification of structures with characteristics different from those of the platform investigated here should be made by numerical studies based on carefully defined climatological and hydrodynamic assumptions. The hydrodynamic assumptions should make allowance for the possibility that, if the Keulegan-Carpenter number is very small, the energy dissipation due to fluid viscosity may in some cases be reflected by values $C_d < 0.1$. It is pointed out that data concerning energy losses in cases where the Reynolds numbers are large (say, of the order of one million or more), and the Keulegan-Carpenter numbers are small (of the order of unity or less), are currently not available; it would appear that in the present state of the art such data cannot be obtained from laboratory tests, and that they should be inferred from full-scale measurements.

Acknowledgment

This work was supported by the Technology Assessment and Research Branch, Minerals Management Service, United States Department of the Interior.

References

1. Simiu, E., and Scanlan, R. H., Wind Effects on Structures, Second Edition, Wiley-Interscience, New York, 1986.

2. Salvesen, N., et al., "Computation of Nonlinear Surge Motions of Tension Leg Platforms," OTC Paper 4394, Proceedings, Offshore Technology Conference, Vol. 4, Houston, TX, May, 1982, pp. 199-216.

3. Simiu, E., and Leigh, S. D., "Turbulent Wind and Tension Leg Platform Surge," Journal of Structural Engineering, April, 1984, pp. 785-802.

4. Sarpkaya, T., "Force on a circular cylinder in viscous oscillatory flow at low Keulegan-Carpenter numbers," Journal of Fluid Mechanics, Vol. 165, April 1986, pp. 61-71.

5. Sarpkaya, T. and Isaacson, M., Mechanics of Wave Forces on Offshore Structures, Van Nostrand Reinhold Company, New

York, 1981.

6. Bearman, P. W., Graham, J. M. R., and Singh, S., "Forces on cylinders in harmonically oscillating flow," in <u>Mechanics of Wave-Induced Forces on Cylinders</u>, T. L. Shaw (ed.), Pitman, San Francisco, 1979, pp. 437-449.

7. Bearman, P. W., et al., "Forces on cylinders in viscous oscillatory flow at low Keulegan-Carpenter numbers," <u>Journal of Fluid Mechanics</u>, Vol. 154, 1985, pp. 337-356.

8. Wang, C.-Y., "On high-frequency oscillatory viscous flows," <u>Journal of Fluid Mechanics</u>, Vol. 32, 1968, pp. 55-86.

9. Honji, H., "Streaked flow around an oscillating circular cylinder," <u>Journal of Fluid Mechanics</u>, Vol. 107, 1981, pp.509 - 520.

10. Hall, P., "On the stability of the unsteady boundary layer on a cylinder oscillating transversely in a viscous fluid," <u>Journal of Fluid Mechanics</u>, Vol. 146, pp.346-367.

11. Salvesen, N., Meinhold, M. J., and Yue, D. K., <u>Nonlinear Motions and Forces on Tension Leg Platforms</u>, USCG-M-84-4 (16718), U.S. Department of Transportation, United States Coast Guard, Washington, D.C., May 1984.

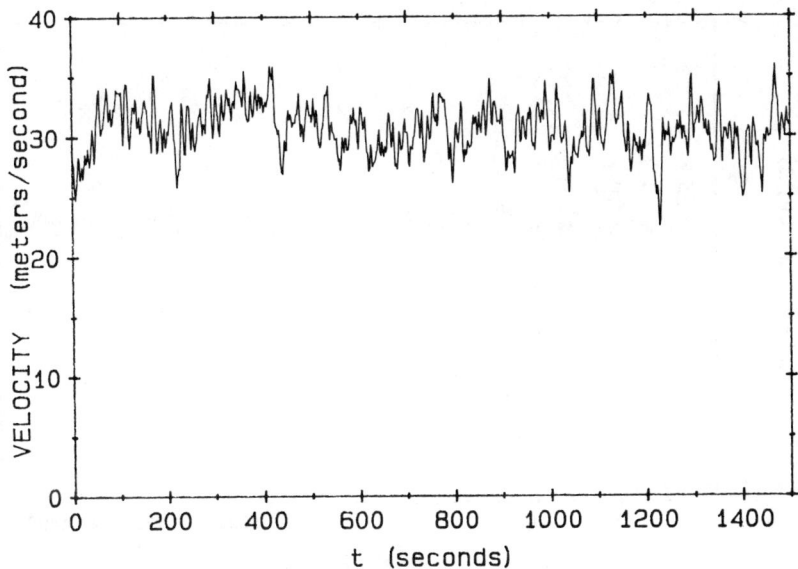

Fig.1 Longitudinal wind velocity fluctuations (Mean speed U(35m) = 30m/s).

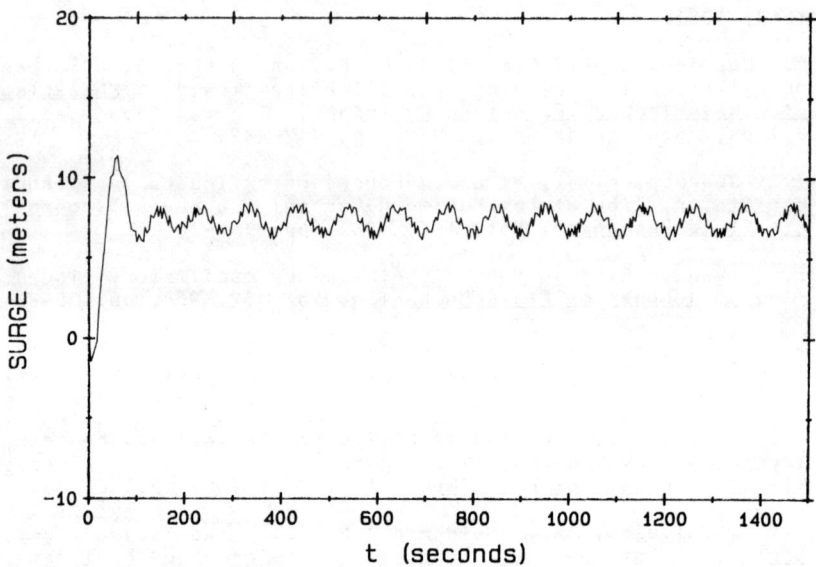

Fig. 2. Surge response. Harmonic wind with mean speed U(35m) = 15m/s; H = 3m; Cd = 0.62; Cm = 1.8 (with current).

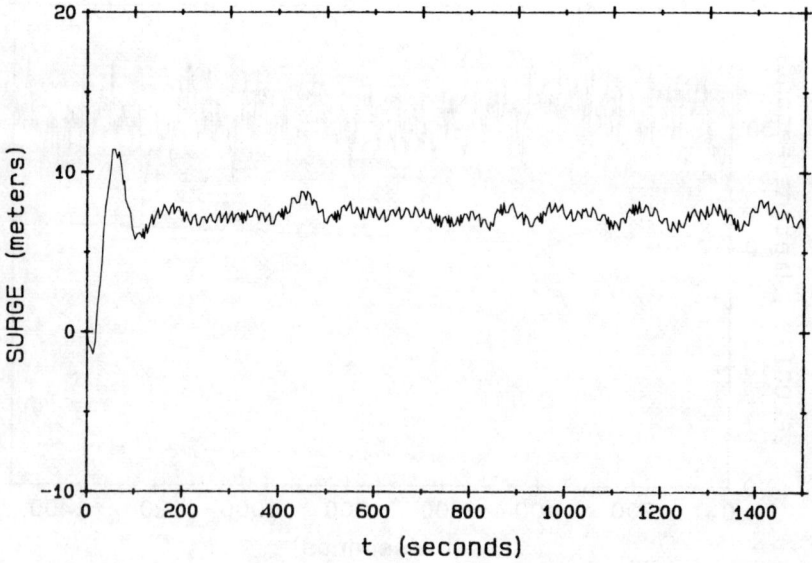

Fig. 3. Surge response. Turbulent wind with mean speed U(35m) = 15m/s; H = 3m; Cd = 0.62; Cm = 1.8 (with current).

WIND EFFECTS AMPLIFICATION

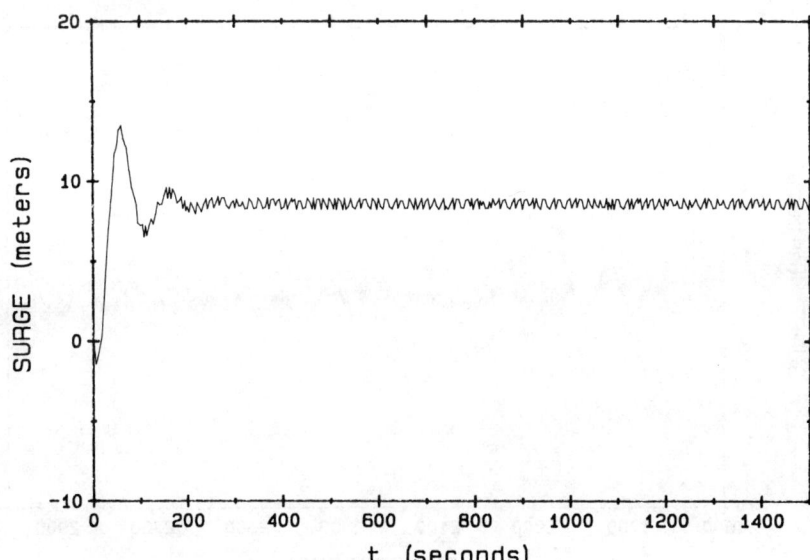

Fig. 4. Surge response. Constant wind with speed $1.25 \times 15 = 18.75$m/s; $H = 3$m; $Cd = 0.62$; $Cm = 1.8$ (with current).

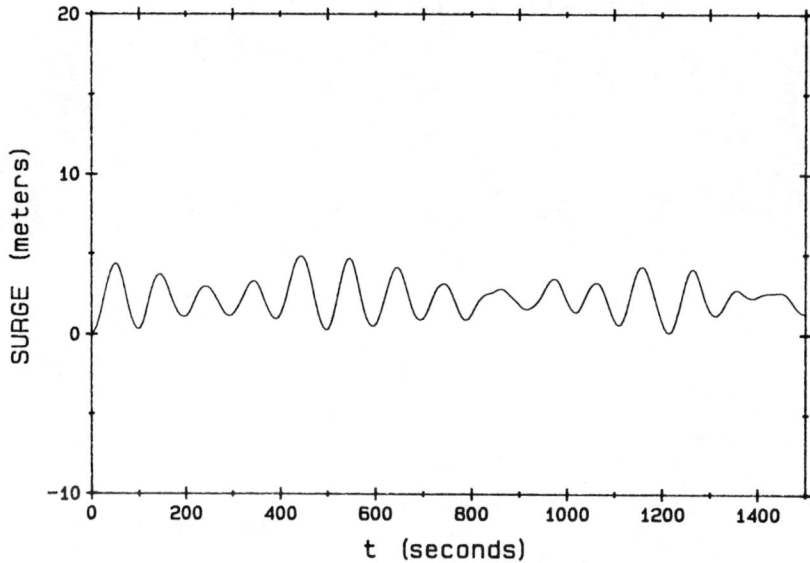

Fig. 5. Surge response. Turbulent wind with mean speed $U(35m) = 15$m/s; $Cd = 0.62$; $Cm = 1.8$ (no waves or current).

70 WIND EFFECTS ON OFFSHORE STRUCTURES

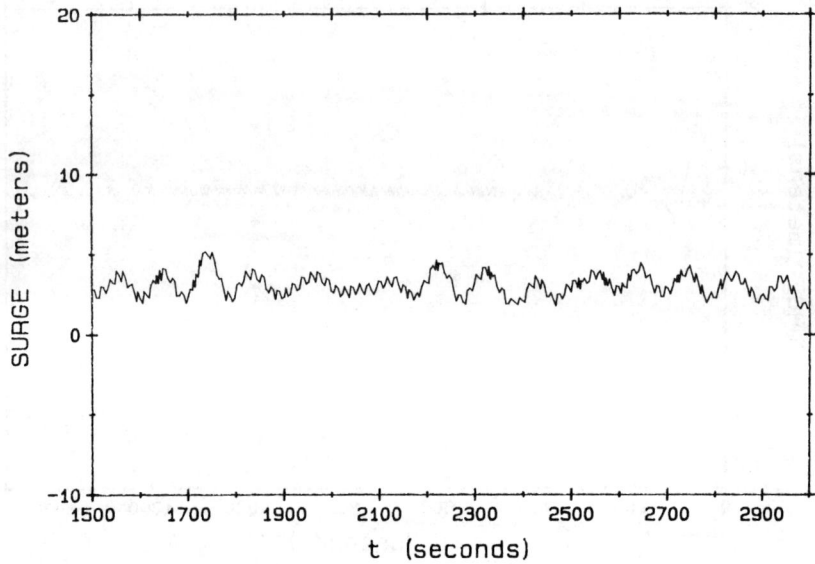

Fig. 6 Surge response. Turbulent wind with mean speed U(35m) = 15m/s; H = 3m; Cd = 0.1; Cm = 1.8 (with current).

SUBJECT INDEX
Page number refers to first page of paper.

Damping, 59
Drag coefficient, 59

Floating structures, 1, 13

Hydrodynamics, 59

Laboratory tests, 43
Load combinations, 55

Offshore platforms, 25
Offshore structures, 1
Offshorre drilling, 25

Simulation models, 43
Stability analysis, 13

Tension leg platforms, 43, 55, 59
Test procedures, 13

Wave forces, 55
Wind loads, 1, 25, 43, 55
Wind tunnel models, 1
Wind tunnel test, 13, 55

AUTHOR INDEX
Page number refers to first page of paper.

Bjerregaard, E. T. D., 13

Cook, Graham R., 59

Davenport, A. G., 55

Finnigan, T., 43
Freathy, P., 25

Hansen, S. O., 13
Helliwell, S., 25

Kareem, A., 43
Kumarasena, Thusitha, 59

Liu, S. L., 43
Lu, P. C., 43

Macha, J. M., 1

Simiu, Emil, 59

Vickery, B. J., 25
Vickery, P. J., 55